Leveled Texts for Mathematics

Algebra and Algebraic Thinking

Author

Lori Barker

Consultant

Barbara Talley, M.S.
Texas A&M University

Publishing Credits

Dona Herweck Rice, *Editor-in-Chief*; Lee Aucoin, *Creative Director*; Don Tran, *Print Production Manager*;
Sara Johnson, M.S.Ed., *Senior Editor*; Hillary Wolfe, *Editor*; Stephanie Paris, *Editor*;
Evelyn Garcia, *Associate Education Editor*; Neri Garcia, *Cover Designer*; Juan Chavolla, *Production Artist*;
Stephanie Reid, *Photo Editor*; Corinne Burton, M.S.Ed., *Publisher*

All images from Shutterstock.com

Shell Education

5301 Oceanus Drive

Huntington Beach, CA 92649

http://www.shelleducation.com

ISBN 978-1-4258-0716-0

©2011 Shell Educational Publishing, Inc.

Reprinted 2013

The classroom teacher may reproduce copies of materials in this book for classroom use only. The reproduction of any part for an entire school or school system is strictly prohibited. No part of this publication may be transmitted, stored, or recorded in any form without written permission from the publisher.

Table of Contents

What Is Differentiation? 4
How to Differentiate Using This Product 5
General Information About the Student Populations 6
    Below-Grade-Level Students 6
    English Language Learners 6
    On-Grade-Level Students 7
    Above-Grade-Level Students 7
Strategies for Using the Leveled Texts 8
    Below-Grade-Level Students 8
    English Language Learners 11
    Above-Grade-Level Students 14
How to Use This Product 16
    Readability Chart 16
    Components of the Product 16
    Tips for Managing the Product 18
    Correlations to Mathematics Standards 19
Leveled Texts 21
    Various Variables 21
    Shaping Up 29
    Sometimes the Change Is Consistent 37
    Sometimes the Change Changes 45
    It's All Organized 53
    Express It Mathematically 61
    Expressing More…Mathematically 69
    Many Ways to Look at It 77
    Adding Some Balance 85
    Keeping the Balance When Taking Away 93
    The Equations Keep Multiplying 101
    Equation Writing 109
    Everything Has a Place 117
    Moving Around 125
    In a Group 133
Appendices 141
    References Cited 141
    Contents of Teacher Resource CD 142

What Is Differentiation?

Over the past few years, classrooms have evolved into diverse pools of learners. Gifted students, English language learners, special needs students, high achievers, underachievers, and average students all come together to learn from one teacher. The teacher is expected to meet their diverse needs in one classroom. It brings back memories of the one-room schoolhouse during early American history. Not too long ago, lessons were designed to be one size fits all. It was thought that students in the same grade level learned in similar ways. Today, we know that viewpoint to be faulty. Students have differing learning styles, come from different cultures, experience a variety of emotions, and have varied interests. For each subject, they also differ in academic readiness. At times, the challenges teachers face can be overwhelming, as they struggle to figure out how to create learning environments that address the differences they find in their students.

What is differentiation? Carol Ann Tomlinson at the University of Virginia says, "Differentiation is simply a teacher attending to the learning needs of a particular student or small group of students, rather than teaching a class as though all individuals in it were basically alike" (2000). Differentiation can be carried out by any teacher who keeps the learners at the forefront of his or her instruction. The effective teacher asks, "What am I going to do to shape instruction to meet the needs of all my learners?" One method or methodology will not reach all students.

Differentiation encompasses what is taught, how it is taught, and the products students create to show what they have learned. When differentiating curriculum, teachers become the organizers of learning opportunities within the classroom environment. These categories are often referred to as content, process, and product.

- **Content:** Differentiating the content means to put more depth into the curriculum through organizing the curriculum concepts and structure of knowledge.
- **Process:** Differentiating the process means using varied instructional techniques and materials to enhance students' learning.
- **Product:** When products are differentiated, cognitive development and the students' abilities to express themselves improve, as they are given different product options.

Teachers should differentiate content, process, and product according to students' characteristics, including students' readiness, learning styles, and interests.

- **Readiness:** If a learning experience aligns closely with students' previous skills and understanding of a topic, they will learn better.
- **Learning styles:** Teachers should create assignments that allow students to complete work according to their personal preferences and styles.
- **Interests:** If a topic sparks excitement in the learners, then students will become involved in learning and better remember what is taught.

How to Differentiate Using This Product

According to the Common Core State Standards (2010), all students need to learn to read and discuss concepts across the content areas in order to be prepared for college and beyond. The leveled texts in this series help teachers differentiate mathematics content for their students to allow all students access to the concepts being explored. Each book has 15 topics, and each topic has a text written at four different reading levels. (See page 17 for more information.) While these texts are written at a variety of reading levels, all the levels remain strong in presenting the mathematics content and vocabulary. Teachers can focus on the same content standard or objective for the whole class, but individual students can access the content at their instructional reading levels rather than at their frustration levels.

Determining your students' instructional reading levels is the first step in the process. It is important to assess their reading abilities often so they do not get tracked into one level. Below are suggested ways to determine students' reading levels.

- **Running records:** While your class is doing independent work, pull your below-grade-level students aside, one at a time. Have them read aloud the lowest level of a text (the star level) individually as you record any errors they make on your own copy of the text. If students read accurately and fluently and comprehend the material, move them up to the next level and repeat the process. Following the reading, ask comprehension questions to assess their understanding of the material. Use your judgment to determine whether students seem frustrated as they read. As a general guideline, students reading below 90% accuracy are likely to feel frustrated as they read. There are also a variety of published reading assessment tools that can be used to assess students' reading levels with the running record format.

- **Refer to other resources:** Other ways to determine instructional reading levels include checking your students' Individualized Education Plans (IEP), asking the school's resource teachers, or reviewing test scores. All of these resources should be able to give you the additional information you need to determine the reading level to begin with for your students.

Teachers can also use the texts in this series to scaffold the content for their students. At the beginning of the year, students at the lowest reading levels may need focused teacher guidance. As the year progresses, teachers can begin giving students multiple levels of the same text to allow them to work independently to improve their comprehension. This means each student would have a copy of the text at his or her independent reading level and instructional reading level. As students read the instructional-level texts, they can use the lower texts to better understand the difficult vocabulary. By scaffolding the content in this way, teachers can support students as they move up through the reading levels. This will encourage students to work with texts that are closer to the grade level at which they will be tested.

General Information About the Student Populations

Below-Grade-Level Students

By Dennis Benjamin

Gone are the days of a separate special education curriculum. Federal government regulations require that special needs students have access to the general education curriculum. For the vast majority of special needs students today, their Individualized Education Plans (IEPs) contain current and targeted performance levels but few short-term content objectives. In other words, the special needs students are required to learn the same content as their on-grade-level peers.

Be well aware of the accommodations and modifications written in students' IEPs. Use them in your teaching and assessment so they become routine. If you hold high expectations of success for all of your students, their efforts and performances will rise as well. Remember the root word of *disability* is *ability*. Go to the root needs of the learner and apply good teaching. The results will astound and please both of you.

English Language Learners

By Marcela von Vacano

Many school districts have chosen the inclusion model to integrate English language learners into mainstream classrooms. This model has its benefits as well as its drawbacks. One benefit is that English language learners may be able to learn from their peers by hearing and using English more frequently. One drawback is that these second-language learners cannot understand academic language and concepts without special instruction. They need sheltered instruction to take the first steps toward mastering English. In an inclusion classroom, the teacher may not have the time or necessary training to provide specialized instruction for these learners.

Acquiring a second language is a lengthy process that integrates listening, speaking, reading, and writing. Students who are newcomers to the English language are not able to process information until they have mastered a certain number of structures and vocabulary words. Students may learn social language in one or two years. However, academic language takes up to eight years for most students.

Teaching academic language requires good planning and effective implementation. Pacing, or the rate at which information is presented, is another important component in this process. English language learners need to hear the same word in context several times, and they need to practice structures to internalize the words. Reviewing and summarizing what was taught are absolutely necessary for English language learners.

General Information About the Student Populations (cont.)

On-Grade-Level Students

By Wendy Conklin

On-grade-level students often get overlooked when planning curriculum. More emphasis is usually placed on those who struggle and, at times, on those who excel. Teachers spend time teaching basic skills and even go below grade level to ensure that all students are up to speed. While this is a noble endeavor and is necessary at times, in the midst of it all, the on-grade-level students can get lost in the shuffle. We must not forget that differentiated strategies are good for the on-grade-level students, too. Providing activities that are too challenging can frustrate these students; on the other hand, assignments that are too easy can be boring and a waste of their time. The key to reaching this population successfully is to find just the right level of activities and questions while keeping a keen eye on their diverse learning styles.

Above-Grade-Level Students

By Wendy Conklin

In recent years, many state and school district budgets have cut funding that has in the past provided resources for their gifted and talented programs. The push and focus of schools nationwide is proficiency. It is important that students have the basic skills to read fluently, solve problems, and grasp mathematical concepts. As a result, funding has been redistributed in hopes of improving test scores on state and national standardized tests. In many cases, the attention has focused only on improving low test scores to the detriment of the gifted students who need to be challenged.

Differentiating the products you require from your students is a very effective and fairly easy way to meet the needs of gifted students. Actually, this simple change to your assignments will benefit all levels of students in your classroom. While some students are strong verbally, others express themselves better through nonlinguistic representation. After reading the texts in this book, students can express their comprehension through different means, such as drawings, plays, songs, skits, or videos. It is important to identify and address different learning styles. By giving more open-ended assignments, you allow for more creativity and diversity in your classroom. These differentiated products can easily be aligned with content standards. To assess these standards, use differentiated rubrics.

Strategies for Using the Leveled Texts

Below-Grade-Level Students

By Dennis Benjamin

Vocabulary Scavenger Hunt

A valuable prereading strategy is a Vocabulary Scavenger Hunt. Students preview the text and highlight unknown words. Students then write the words on specially divided pages. The pages are divided into quarters with the following headings: *Definition*, *Sentence*, *Examples*, and *Nonexamples*. A section called *Picture* can be placed over the middle of the chart to give students a visual reminder of the word and its definition.

Example Vocabulary Scavenger Hunt

estimate

Definition	Sentence
to determine roughly the size or quantity	We need to estimate how much paint we will need to buy.
Examples	**Nonexamples**
estimating a distance; estimating an amount	measuring to an exact number; weighing and recording an exact weight

This encounter with new vocabulary enables students to use it properly. The definition identifies the word's meaning in student-friendly language. The sentence should be written so that the word is used in context. This helps the student make connections with background knowledge. Illustrating the sentence gives a visual clue. Examples help students prepare for factual questions from the teacher or on standardized assessments. Nonexamples help students prepare for ***not*** and ***except for*** test questions such as "All of these are polygons *except for*…" and "Which of these terms in this expression is not a constant?" Any information the student was unable to record before reading can be added after reading the text.

Strategies for Using the Leveled Texts (cont.)

Below-Grade-Level Students (cont.)

Graphic Organizers to Find Similarities and Differences

Setting a purpose for reading content focuses the learner. One purpose for reading can be to identify similarities and differences. This is a skill that must be directly taught, modeled, and applied. The authors of *Classroom Instruction That Works* state that identifying similarities and differences "might be considered the core of all learning" (Marzano, Pickering, and Pollock 2001). Higher-level tasks include comparing and classifying information and using metaphors and analogies. One way to scaffold these skills is through the use of graphic organizers, which help students focus on the essential information and organize their thoughts.

Example Classifying Graphic Organizer

Equation	Constants	Variables	Number of Terms
$7 + 12 = 19$	7, 12, 19	none	3
$3x = 12$	12	x	2
$a + b$	none	a, b	2
$a^2 + b^2 + c^2$	none	a, b, c	3

The Riddles Graphic Organizer allows students to compare and contrast two-dimensional shapes using riddles. Students first complete a chart you've designed. Then, using that chart, they can write summary sentences. They do this by using the riddle clues and reading across the chart. Students can also read down the chart and write summary sentences. With the chart below, students could write the following sentences: A circle is not a polygon. The interior angles of a triangle always add up to 180°.

Example Riddles Graphic Organizer

What Am I?	Circle	Square	Triangle	Rectangle
I come in different configurations.			X	X
I am a polygon.		X	X	X
I am a closed shape.	X	X	X	X
My interior angles always add up to 180°.			X	
I have at least three vertices.		X	X	X

Strategies for Using the Leveled Texts (cont.)

Below-Grade-Level Students (cont.)

Framed Outline

This is an underused technique that bears great results. Many below-grade-level students have problems with reading comprehension. They need a framework to help them attack the text and gain confidence in comprehending the material. Once students gain confidence and learn how to locate factual information, the teacher can fade out this technique.

There are two steps to successfully using this technique. First, the teacher writes cloze sentences. Second, the students complete the cloze activity and write summary sentences.

Example Framed Outline

A _____ graph is used to show how two variables may be related to each other. A _____ graph shows the relationship between dependent and independent variables.

Summary Sentences:

A good graph should correctly show the data. It should have a title; the axes should be labeled correctly; it will should show the proper scale.

Modeling Written Responses

A frequent criticism heard by special educators is that below-grade-level students write poor responses to content-area questions. This problem can be remedied if resource and classroom teachers model what good answers look like. While this may seem like common sense, few teachers take the time to do this. They just assume all children know how to respond in writing.

This is a technique you may want to use before asking your students to respond to the You Try It questions associated with the leveled texts in this series. First, read the question aloud. Then, write the question on the board or an overhead and think aloud about how you would go about answering the question. Next, solve the problem showing all the steps. Introduce the other problems and repeat the procedure. Have students explain how they solved the problems in writing so that they make the connection that quality written responses are part of your expectations.

Strategies for Using the Leveled Texts (cont.)

English Language Learners

By Marcela von Vacano

Effective teaching for English language learners requires effective planning. In order to achieve success, teachers need to understand and use a conceptual framework to help them plan lessons and units. There are six major components to any framework. Each is described in detail below.

1. Select and Define Concepts and Language Objectives—Before having students read one of the texts in this book, the teacher must first choose a mathematical concept and language objective (reading, writing, listening, or speaking) appropriate for the grade level. Then, the next step is to clearly define the concept to be taught. This requires knowledge of the subject matter, alignment with local and state objectives, and careful formulation of a statement that defines the concept. This concept represents the overarching idea. The mathematical concept should be written on a piece of paper and posted in a visible place in the classroom.

By the definition of the concept, post a set of key language objectives. Based on the content and language objectives, select essential vocabulary from the text. The number of new words selected should be based on students' English language levels. Post these words on a word wall that may be arranged alphabetically or by themes.

2. Build Background Knowledge—Some English language learners may have a lot of knowledge in their native language, while others may have little or no knowledge. The teacher will want to build the background knowledge of the students using different strategies such as the following:

> **Visuals:** Use posters, photographs, postcards, newspapers, magazines, drawings, and video clips of the topic you are presenting.
>
> **Realia:** Bring real-life objects to the classroom. If you are teaching about measurement, bring in items such as thermometers, scales, time pieces, and rulers.
>
> **Vocabulary and Word Wall:** Introduce key vocabulary in context. Create families of words. Have students draw pictures that illustrate the words and write sentences about the words. Also, be sure you have posted the words on a word wall in your classroom.
>
> **Desk Dictionaries:** Have students create their own desk dictionaries using index cards. On one side, they should draw a picture of the word. On the opposite side, they should write the word in their own language and in English.

Strategies for Using the Leveled Texts (cont.)

English Language Learners (cont.)

3. **Teach Concepts and Language Objectives**—The teacher must present content and language objectives clearly. He or she must engage students using a hook and must pace the delivery of instruction, taking into consideration students' English language levels. The concept or concepts to be taught must be stated clearly. Use the first languages of the students whenever possible or assign other students who speak the same languages to mentor and to work cooperatively with the English language learners.

   Lev Semenovich Vygotsky, a Russian psychologist, wrote about the Zone of Proximal Development (ZPD). This theory states that good instruction must fill the gap that exists between the present knowledge of a child and the child's potential (1978). Scaffolding instruction is an important component when planning and teaching lessons. English language learners cannot jump stages of language and content development. You must determine where the students are in the learning process and teach to the next level using several small steps to get to the desired outcome. With the leveled texts in this series and periodic assessment of students' language levels, teachers can support students as they climb the academic ladder.

4. **Practice Concepts and Language Objectives**—English language learners need to practice what they learn with engaging activities. Most people retain knowledge best after applying what they learn to their own lives. This is definitely true for English language learners. Students can apply content and language knowledge by creating projects, stories, skits, poems, or artifacts that show what they learned. Some activities should be geared to the right side of the brain. For students who are left-brain dominant, activities such as defining words and concepts, using graphic organizers, and explaining procedures should be developed. The following teaching strategies are effective in helping students practice both language and content:

   Simulations: Students learn by doing. For example, when teaching about data analysis, have students do a survey about their classmates' favorite sports. First, students make a list of questions and collect the necessary data. Then, they tally the responses and determine the best way to represent the data. Lastly, students create a graph that shows their results and display it in the classroom.

   Literature response: Read a text from this book. Have students choose two concepts described or introduced in the text. Ask students to create a conversation two people might have to debate which concept is useful. Or, have students write journal entries about real-life ways they use these mathematical concepts.

Strategies for Using the Leveled Texts (cont.)

English Language Learners (cont.)

4. Practice Concepts and Language Objectives (cont.)

Have a short debate: Make a controversial statement such as "It isn't necessary to learn addition." After reading a text in this book, have students think about the question and take a position. As students present their ideas, one student can act as a moderator.

Interview: Students may interview a member of the family or a neighbor in order to obtain information regarding a topic from the texts in this book. For example: What are some ways you use geometry in your work?

5. Evaluation and Alternative Assessments—
We know that evaluation is used to inform instruction. Students must have the opportunity to show their understanding of concepts in different ways and not only through standard assessments. Use both formative and summative assessments to ensure that you are effectively meeting your content and language objectives. Formative assessment is used to plan effective lessons for a particular group of students. Summative assessment is used to find out how much the students have learned. Other authentic assessments that show day-to-day progress are: text retelling, teacher rating scales, student self-evaluations, cloze testing, holistic scoring of writing samples, performance assessments, and portfolios. Periodically assessing student learning will help you ensure that students continue to receive the correct levels of texts.

6. Home-School Connection—
The home-school connection is an important component in the learning process for English language learners. Parents are the first teachers, and they establish expectations for their children. These expectations help shape the behavior of their children. By asking parents to be active participants in the education of their children, students get a double dose of support and encouragement. As a result, families become partners in the education of their children and chances for success in your classroom increase.

You can send home copies of the texts in this series for parents to read with their children. You can even send multiple levels to meet the needs of your second-language parents as well as your students. In this way, you are sharing your mathematics content standards with your whole second-language community.

Strategies for Using the Leveled Texts (cont.)

Above-Grade-Level Students

By Wendy Conklin

Open-Ended Questions and Activities

Teachers need to be aware of activities that provide a ceiling that is too low for gifted students. When given activities like this, gifted students become bored. We know these students can do more, but how much more? Offering open-ended questions and activities will give high-ability students the opportunities to perform at or above their ability levels. For example, ask students to evaluate mathematical topics described in the texts, with questions such as "Do you think students should be allowed to use calculators in math?" or "What do you think you would need to build a two-story dog house?" These questions require students to form opinions, think deeply about the issues, and form several different responses in their minds. To questions like these, there really is no single correct answer.

The generic, open-ended question stems listed below can be adapted to any topic. There is one You Try It question for each topic in this book. Use questions or statements like the ones shown here to develop further discussion for the leveled texts.

- In what ways did…
- How might you have done this differently…
- What if…
- What are some possible explanations for…
- How does this affect…
- Explain several reasons why…
- What problems does this create…
- Describe the ways…
- What is the best…
- What is the worst…
- What is the likelihood…
- Predict the outcome…
- Form a hypothesis…
- What are three ways to classify…
- Support your reason…
- Make a plan for…
- Propose a solution…
- What is an alternative to…

Strategies for Using the Leveled Texts (cont.)

Above-Grade-Level Students (cont.)

Student-Directed Learning

Because they are academically advanced, above-grade-level students are often the leaders in classrooms. They are more self-sufficient learners, too. As a result, there are some student-directed strategies that teachers can employ successfully with these students. Remember to use the texts in this book as jump starts so that students will be interested in finding out more about the mathematical concepts presented. Above-grade-level students may enjoy any of the following activities:

- Writing their own questions, exchanging their questions with others, and grading the responses.
- Reviewing the lesson and teaching the topic to another group of students.
- Reading other nonfiction texts about these mathematical concepts to further expand their knowledge.
- Writing the quizzes and tests to go along with the texts.
- Creating illustrated timelines to be displayed as visuals for the entire class.
- Putting together multimedia presentations about the mathematical concepts.

Tiered Assignments

Teachers can differentiate lessons by using tiered assignments, or scaffolded lessons. Tiered assignments are parallel tasks designed to have varied levels of depth, complexity, and abstractness. All students work toward one goal, concept, or outcome, but the lesson is tiered to allow for different levels of readiness and performance. As students work, they build on their prior knowledge and understanding. Students are motivated to be successful according to their own readiness and learning preferences.

Guidelines for writing tiered lessons include the following:

1. Pick the skill, concept, or generalization that needs to be learned.
2. Think of an on-grade-level activity that teaches this skill, concept, or generalization.
3. Assess the students using classroom discussions, quizzes, tests, or journal entries and place them in groups.
4. Take another look at the activity from Step 2. Modify this activity to meet the needs of the below-grade-level and above-grade-level learners in the class. Add complexity and depth for the above-grade-level students. Add vocabulary support and concrete examples for the below-grade-level students.

How to Use This Product

Readability Chart

Title of the Text	Star	Circle	Square	Triangle
Various Variables	1.7	3.4	5.0	6.5
Shaping Up	2.0	3.1	5.0	6.5
Sometimes the Change Is Consistent	2.2	3.0	5.0	6.5
Sometimes the Change Changes	2.0	3.4	5.0	6.6
It's All Organized	2.2	3.4	5.0	6.5
Express It Mathematically	2.2	3.0	5.0	6.7
Expressing More…Mathematically	1.9	3.3	5.0	6.5
Many Ways to Look at It	2.2	3.2	5.0	6.6
Adding Some Balance	1.7	3.5	5.1	6.5
Keeping the Balance When Taking Away	1.8	3.0	5.0	6.5
The Equations Keep Multiplying	2.2	3.3	5.0	6.5
Equation Writing	2.2	3.4	5.0	6.5
Everything Has a Place	1.9	3.1	5.3	6.5
Moving Around	2.1	3.5	5.1	6.6
In a Group	2.1	3.2	5.5	6.6

Components of the Product

Strong Image Support

- Each level of text includes important visual support. These images, diagrams, photographs, and illustrations add interest to the texts and help students comprehend the mathematical concepts. The images also serve as visual support for second-language learners. They make the texts more context-rich and bring the examples to life.

How to Use This Product (cont.)

Components of the Product (cont.)

Comprehension Questions

- The introduction often includes a challenging question or riddle. The answer can be found on the next page at the end of the lesson.
- Each level of text includes a You Try It section where the students are asked to solve problems using the skill or concept discussed in the text.
- Although the mathematics is the same, the questions may be worded slightly differently depending on the reading level of the passage.

The Levels

- There are 15 topics in this book. Each topic is leveled to four different reading levels. The images and fonts used for each level within a topic look the same.
- Behind each page number, you'll see a shape. These shapes indicate the reading levels of each text so that you can make sure students are working with the correct texts. The reading levels fall into the ranges indicated below. See the chart on page 16 for the specific reading levels of each lesson.

Leveling Process

- The texts in this series were originally authored by mathematics educators. A reading expert went through the texts and leveled each one to create four distinct reading levels.
- A mathematics expert then reviewed each passage for accuracy and mathematical language.
- The texts were then leveled one final time to ensure the editorial changes made during the process kept them within the ranges described to the left.

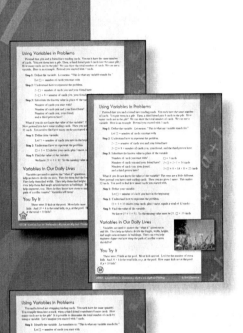

Levels 1.5–2.2

Levels 3.0–3.5

Levels 5.0–5.5

Levels 6.5–7.2

How to Use This Product (cont.)

Tips for Managing the Product

How to Prepare the Texts

- When you copy these texts, be sure you set your copier to copy photographs. Run a few test pages and adjust the contrast as necessary. If you want the students to be able to appreciate the images, you need to carefully prepare the texts for them.

- You also have full-color versions of the texts provided in PDF form on the CD. (See page 142 for more information.) Depending on how many copies you need to make, printing the full-color versions and copying those might work best for you.

- Keep in mind that you should copy two-sided to two-sided if you pull the pages out of the book. The shapes behind the page numbers will help you keep the pages organized as you prepare them.

Distributing the Texts

Some teachers wonder about how to hand the texts out within one classroom. They worry that students will feel insulted if they do not get the same papers as their neighbors. The first step in dealing with these texts is to set up your classroom as a place where all students learn at their individual instructional levels. Making this clear as a fact of life in your classroom is key. Otherwise, the students may constantly ask about why their work is different. You do not need to get into the technicalities of the reading levels. Just state it as a fact that every student will not be working on the same assignment every day. If you do this, then passing out the varied levels is not a problem. Just pass them to the correct students as you circle the room.

If you would rather not have students openly aware of the differences in the texts, you can try these ways to pass out the materials:

- Make a pile in your hands from star to triangle. Put your finger between the circle and square levels. As you approach each student, you pull from the top (star), above your finger (circle), below your finger (square), or the bottom (triangle). If you do not hesitate too much in front of each desk, the students will probably not notice.

- Begin the class period with an opening activity. Put the texts in different places around the room. As students work quietly, circulate and direct students to the right locations for retrieving the texts you want them to use.

- Organize the texts in small piles by seating arrangement so that when you arrive at a group of desks you have just the levels you need.

How to Use This Product (cont.)

Correlation to Mathematics Standards

Shell Education is committed to producing educational materials that are research and standards based. In this effort, we have correlated all of our products to the academic standards of all 50 United States, the District of Columbia, the Department of Defense Dependent Schools, and all Canadian provinces. We have also correlated to the Common Core State Standards.

How to Find Standards Correlations

To print a customized correlation report of this product for your state, visit our website at **http://www.shelleducation.com** and follow the on-screen directions. If you require assistance in printing correlation reports, please contact Customer Service at 1-877-777-3450.

Purpose and Intent of Standards

Legislation mandates that all states adopt academic standards that identify the skills students will learn in kindergarten through grade twelve. Many states also have standards for Pre-K. This same legislation sets requirements to ensure the standards are detailed and comprehensive.

Standards are designed to focus instruction and guide adoption of curricula. Standards are statements that describe the criteria necessary for students to meet specific academic goals. They define the knowledge, skills, and content students should acquire at each level. Standards are also used to develop standardized tests to evaluate students' academic progress. Teachers are required to demonstrate how their lessons meet state standards. State standards are used in the development of all of our products, so educators can be assured they meet the academic requirements of each state.

TESOL Standards

The lessons in this book promote English language development for English language learner. The standards listed on the Teacher Resource CD support the language objectives presented throughout the lessons.

NCTM Standards Correlation Chart

The chart on the next page shows the correlation to the National Council for Teachers of Mathematics (NCTM) standards. This chart is also available on the Teacher Resource CD (nctm.pdf).

NCTM Standards

NCTM Standard	Lesson	Page
Describe, extend, and make generalizations about geometric and numeric patterns	Shaping Up; Sometimes the Change Is Consistent; Sometimes the Change Changes; It's All Organized	29–60
Represent and analyze patterns and functions, using words, tables, and graphs	Everything Has a Place; Moving Around; In a Group	117–140
Identify such properties as commutativity, associativity, and distributivity and use them to compute with whole numbers	Adding Some Balance; Keeping the Balance When Taking Away; The Equations Keep Multiplying; Moving Around; In a Group	85–108, 125–140
Represent the idea of a variable as an unknown quantity using a letter or a symbol	Various Variables; Express It Mathematically; Expressing More...Mathematically; Adding Some Balance; The Equations Keep Multiplying; Equation Writing; Everything Has a Place	21–28, 61–76, 85–92, 101–124
Express mathematical relationships using equations	Various Variables; The Equations Keep Multiplying; Many Ways to Look at It; Equation Writing	21–28, 101–116
Model problem situations with objects and use representations such as graphs, tables, and equations to draw conclusions	Various Variables; The Equations Keep Multiplying; Many Ways to Look at It; Equation Writing	21–28, 77–84, 101–116
Investigate how a change in one variable relates to a change in a second variable	Many Ways to Look at It; Adding Some Balance; Keeping the Balance When Taking Away; Moving Around; In a Group	77–100, 125–140
Identify and describe situations with constant or varying rates of change and compare them	Sometimes the Change is Consistent; Sometimes the Change Changes	37–52

Standards are listed with the permission of the National Council of Teachers of Mathematics (NCTM). NCTM does not endorse the content or validity of these alignments.

Various Variables

Think of picking up a dog. It weighs 60 lbs. on Earth. Did you know that the same dog weighs less on the moon? Gravity is less on the moon. It is 6 times less than on Earth. The equation $w \div 6$ can show the dog's weight on the moon. Replace the w with 60. Then divide. $60 \div 6 = 10$. So, we know that the dog only weighs 10 lbs. on the moon!

Basic Facts

A **variable** is a symbol. It is used to stand for a number. It can be a letter. It can be a picture. Or it can be any other symbol. Think of $5 + \_\_ = 7$.

5 plus 2 equals 7. So, we can fill in the blank with a 2. We can write $5 + \underline{2} = 7$.

Type of variable	Example	When is this example true?
Drawing or picture	$5 + \square = 7$	When $\square = 2$
Letter or symbol	$5 + n = 7$	When $n = 2$

Any symbol can be a variable. But, try not to use some letters. They can confuse people. Think about the letter o. It might look like a zero. A lowercase l can look like the number 1. Think of multiplication. Avoid using x as your variable. People might think x means "times" or "multiply"!

Here are three other ways of writing multiplication. They will help keep things clear. Each one means "2 times a":

A raised dot	$2 \cdot a$
Parentheses	$2(a)$
Writing a number next to the variable	$2a$

Values of Variables

The value of a variable is not always the same. This is why it is called a variable. Its value can vary. That means it can change.

Think of $2 + \square = 5$. What is the value of $\square$? It has a value of 3.
Now think of $1 + \square = 3$. What is the value of $\square$ here? It has a value of 2.

Pretend you need milk cartons. You need them for a class project. For one class, you need 20 milk cartons. What if you need more? What if you need them for many classes? We can let $\square$ stand for the number of classes. We can write $20 \cdot \square$ for the total number of milk cartons. Here, "$20 \cdot \square$" means "20 times the number of classes."

Using Variables in Problems

Pretend that you and a friend have trading cards. You each have the same number of cards. You put them into a pile. Then, a third friend puts 8 cards into the same pile. How many cards are in the pile? We can show the total number of cards. We can use a variable. Here is an example. Pretend you started with 7 cards.

Step 1: Define the variable. *Let* means: "This is what my variable stands for."

　Let □ = number of cards you start with

Step 2: Understand how to represent the problem.

　2 • □ = number of cards you and your friend have

　2 • □ + 8 = number of cards you, your friend, and the third person have

Step 3: Substitute the known value in place of the variable.

Number of cards you start with?	□ = 7 cards
Number of cards you and your friend have?	2 • □ = 2 • 7 = 14 cards
Number of cards you, your friend, and a third person have?	2 • □ + 8 = 14 + 8 = 22 cards

What if you do not know the value of the variable? The steps are a little different. Now pretend you have some trading cards. Then you are given 5 more. This makes 32 cards. You need to find how many cards you started with.

Step 1: Define your variable.

　Let □ = number of cards you have in the beginning

Step 2: Understand how to represent the problem.

　□ + 5 = 32 (shows your cards, plus 5 more, equals a total of 32 cards)

Step 3: Find the value of the variable.

　We know 27 + 5 = 32. So the missing value must be 27. □ = 27 cards

Variables in Our Daily Lives

Variables are used to answer the "what if" questions. They help architects decide on sizes. They let them find the length. They help them find width. They help them find height. They even help them find angle measurements in buildings. They can help engineers, too. How do they know how steep to make the path of a roller coaster? Variables tell them!

You Try It

There were 20 kids at the pool. More kids came. Let *k* be the number of new kids. And 20 + *k* is the total kids, or *p*, at the pool. How many kids are at the pool if the total = 35 kids?

Various Variables

Imagine you can pick up a 60 lb. dog on Earth. Did you know that the same dog weighs less on the moon? Gravity is six times less on the moon than Earth. The expression $w \div 6$ can show the dog's weight on the moon. Replace the w with 60 and divide. Since $60 \div 6$ is 10, we can figure out that the dog only weighs 10 lbs. on the moon!

Basic Facts

A **variable** is a symbol that is used to stand for a number. It can be a letter. It can be a picture. Or it can be any other symbol. Think of $5 + \underline{} = 7$.

Since 5 plus 2 equals 7, we can fill in the blank with a 2 and write $5 + \underline{2} = 7$.

Type of variable	Example	When is this example true?
Drawing or picture	$5 + \square = 7$	When $\square = 2$
Letter or symbol	$5 + n = 7$	When $n = 2$

Most any symbol can be used as a variable. But, try to avoid using some letters. This is because they can lead to confusion. For example, the letter o might look like a zero. A lowercase l can look like the number 1. And, if you are doing a multiplication problem, avoid using x as your variable. People might think x means "times" or "multiply"!

Here are three other ways of writing multiplication. They will help avoid confusion. Each of these three expressions mean "2 times a":

A raised dot	$2 \cdot a$
Parentheses	$2(a)$
Writing a number next to the variable	$2a$

Values of Variables

The value of a variable is not always the same. This is why it is called a variable. Its value can vary. That means it can change.

Think of $2 + \square = 5$. In this case, the variable $\square$ has a value of 3.
Now think of $1 + \square = 3$. In this case, the variable $\square$ has a value of 2.

Pretend you need milk cartons for a class project. For one class, you need 20 milk cartons. But, what if you must collect for many classes? If we let $\square$ stand for the number of classes, we can write $20 \cdot \square$ for the total number of milk cartons. This is because "$20 \cdot \square$" means "20 times the number of classes."

Using Variables in Problems

Pretend that you and a friend have trading cards. You each have the same number of cards. You put them in a pile. Then, a third friend puts 8 cards in the pile. How many cards are in the pile? We can show the total number of cards. We can use a variable. Here is an example. Pretend you started with 7 cards.

Step 1: Define the variable. *Let* means: "This is what my variable stands for."

Let □ = number of cards you start with

Step 2: Understand how to represent the problem.

2 • □ = number of cards you and your friend have

2 • □ + 8 = number of cards you, your friend, and the third person have

Step 3: Substitute the known value in place of the variable.

Number of cards you start with?   □ = 7 cards

Number of cards you and your friend have?   2 • □ = 2 • 7 = 14 cards

Number of cards you, your friend, and a third person have?   2 • □ + 8 = 14 + 8 = 22 cards

What if you do not know the value of the variable? The steps are a little different. Now pretend you have some trading cards. Then you are given 5 more. This makes 32 cards. You need to find how many cards you started with.

Step 1: Define your variable.

Let □ = number of cards you have in the beginning

Step 2: Understand how to represent the problem.

□ + 5 = 32 (shows your cards, plus 5 more, equals a total of 32 cards)

Step 3: Find the value of the variable.

We know 27 + 5 = 32. So the missing value must be 27.   □ = 27 cards

Variables in Our Daily Lives

Variables are used to answer the "what if" questions in real life. They help architects decide the length, width, height, and angle measurements in buildings. They can even help engineers figure out how steep the path of a roller coaster should be!

You Try It

There were 20 kids at the pool. More kids arrived. Let *k* be the number of extra kids. And 20 + *k* is the total kids, or *p*, at the pool. How many kids are at the pool if *p* = 35 kids?

Various Variables

Could you lift a 60-pound dog on Earth? You probably could on the moon, because gravity's pull is six times less on the moon than on Earth. Use the expression $w \div 6$ to express the dog's weight on the moon. Replace the w with 60 and divide. Since $60 \div 6$ is 10, we can figure out that the dog weighs 10 pounds on the moon!

Basic Facts

A **variable** is a symbol that is used to represent a number. It can be a letter, picture, or any other symbol. Think of $5 + \underline{} = 7$.

Since 5 plus 2 equals 7, we can substitute 2 in the blank and write $5 + \underline{2} = 7$.

Type of variable	Example	When is this example true?
Drawing or picture	$5 + \square = 7$	When $\square = 2$
Letter or symbol	$5 + n = 7$	When $n = 2$

Most any symbol can be used as a variable, but it is advisable to avoid using certain letters, because their similarities to numerical expressions can generate confusion. For example, the letter o might be confused with a zero, and a lowercase l can look like the number 1. And, in a multiplication problem, avoid using x as your variable if you are using × to mean "times" or "multiply"!

Luckily, there are three other ways we can represent multiplication. Each of these three expressions means "2 times a":

A raised dot	$2 \cdot a$
Parentheses	$2(a)$
Writing a number next to the variable	$2a$

Values of Variables

The value of a variable is not always constant, which is why it is called a variable. Its value can vary, which means it can change.

Think of $2 + \square = 5$. In this case, the variable $\square$ has a value of 3.
Now think of $1 + \square = 3$. In this case, the variable $\square$ has a value of 2.

Imagine needing milk cartons for a class project. For one class, you need 20 milk cartons. To collect for many classes, though, how do you determine the total amount of needed cartons? Let $\square$ stand for the number of classes. We can write $20 \cdot \square$ for the total number of milk cartons: "$20 \cdot \square$" means "20 times the number of classes."

Using Variables in Problems

You and a friend are swapping trading cards. You each have the same quantity. You compile them into a stack, when a third friend contributes 8 more cards. How many cards are in the pile? It is possible to determine the total number of cards by using a variable. Let's imagine you started with 7 cards.

Step 1: Identify the variable. *Let* translates to: "This is what my variable stands for."

  Let □ = number of cards you start with

Step 2: Understand how to represent the problem.

  2 • □ = number of cards you and your friend have

  2 • □ + 8 = number of cards you, your friend, and the third person have

Step 3: Substitute the known value in place of the variable.

  Number of cards you start with?   □ = 7 cards

  Number of cards you and your friend have?   2 • □ = 2 • 7 = 14 cards

  Number of cards you, your friend, and a third person have?   2 • □ + 8 = 14 + 8 = 22 cards

What if you do not know the value of the variable? The steps are a little different. Now pretend you have some trading cards. Then you are given 5 more. This makes 32 cards. You need to find how many cards you started with.

Step 1: Define your variable.

  Let □ = number of cards you have in the beginning

Step 2: Understand how to represent the problem.

  □ + 5 = 32 (shows your cards, plus 5 more, equals a total of 32 cards)

Step 3: Find the value of the variable.

  We know 27 + 5 = 32. So the missing value must be 27.   □ = 27 cards

Variables in Our Daily Lives

Variables are used to answer the "what if" questions in real life. They help architects decide the length, width, height, and angle measurements in buildings. They can even help engineers determine how steep the incline of a roller coaster should be!

You Try It

There were 20 kids at the pool. More kids arrived. Let k be the number of extra kids. And $20 + k$ is the total kids, or p, at the pool. How many kids are at the pool if $p = 35$ kids?

Various Variables

Imagine hoisting a 60-pound dog over your head. This may be laborious on Earth, but be relatively effortless on the moon. That's because the dog weighs less on the moon since gravity's pull is six times less effective. Algebraically, the expression $w \div 6$ can represent the dog's weight on the moon. Substitute the w with 60 and divide. Since $60 \div 6$ is 10, it logically follows that the dog conceivably weighs 10 pounds on the moon!

Basic Facts

A variable is any symbol that is used to represent a number in an algebraic expression. A variable can be a letter, picture, or symbol. Think of $5 + \_\_ = 7$.

Since 5 plus 2 equals 7, we can insert a 2 into the blank and write $5 + \underline{2} = 7$.

Type of variable	Example	When is this example true?
Drawing or picture	$5 + \square = 7$	When $\square = 2$
Letter or symbol	$5 + n = 7$	When $n = 2$

Practically any symbol can be used as a variable, but it's judicious to avoid using certain specific letters because their similarities to mathematical expressions can be confusing. The letter o might be confused with a zero, and a lowercase l can resemble the number 1. And, when solving a multiplication problem, avoid x as your variable if × will mean "times" or "multiply"!

To minimize confusion, there are three other ways multiplication symbols can be employed. Each of these three expressions means "2 times a":

A raised dot	$2 \cdot a$
Parentheses	$2(a)$
Writing a number next to the variable	$2a$

Values of Variables

The value of a variable is inconsistent, which is why it is referred to as a variable. Its value can vary, which means it can change.

Think of $2 + \square = 5$. In this case, the variable $\square$ has a value of 3.
Now think of $1 + \square = 3$. In this case, the variable $\square$ has a value of 2.

You need milk cartons for a class project. For one class, you need 20 cartons; you collect for several classes. If $\square$ stands for the number of classes, we write $20 \cdot \square$ for the total number of cartons because "$20 \cdot \square$" means "20 times the number of classes."

Using Variables in Problems

Say you and a friend regularly barter your trading cards. You each have the same quantity. You stack the cards in a pile. Then, a third friend contributes 8 cards to the heap. How many cards are in the pile now? We can show the total number of cards by using a variable. Let's look at an example. Pretend you started with 7 cards.

Step 1: Identify the variable. *Let* translates to: "This is what my variable stands for."

Let □ = number of cards you start with

Step 2: Understand how to represent the problem.

2 • □ = number of cards you and your friend have

2 • □ + 8 = number of cards you, your friend, and the third person have

Step 3: Substitute the known value in place of the variable.

Number of cards you start with?	□ = 7 cards
Number of cards you and your friend have?	2 • □ = 2 • 7 = 14 cards
Number of cards you, your friend, and a third person have?	2 • □ + 8 = 14 + 8 = 22 cards

If the value of the variable is undetermined, the steps become a little more complicated. Pretend you have an undetermined amount of trading cards and you accumulate 5 more, for a total of 32 cards. Could you discern how many cards you started with?

Step 1: Define your variable.

Let □ = number of cards you have in the beginning

Step 2: Understand how to represent the problem.

□ + 5 = 32 (shows your cards, plus 5 more, equals a total of 32 cards)

Step 3: Find the value of the variable.

We know 27 + 5 = 32. So the missing value must be 27. □ = 27 cards

Variables in Our Daily Lives

Variables are invaluable to answer real life "what if" questions. They help architects decide the length, width, height, and angle measurements for constructing buildings. Variables are imperative to engineers as they figure out the degree of incline of a roller coaster's daring track!

You Try It

There were 20 kids splashing at the pool. Suddenly more kids arrived. If k is the number of extra kids, and $20 + k$ is the the new total amount of kids, or p, at the pool, how many kids are at the pool if $p = 35$ kids?

Shaping Up

Look at the cats below. Which cat comes next?

Did you guess a cat with its back arched? "What comes next?" is asked a lot in math. This question can lead to useful thoughts. Sometimes it just leads to a pretty picture.

Basic Facts

A **geometric pattern** uses lines. It uses curves. Or it uses shapes. These patterns must repeat in an orderly way. The figures below make a geometric pattern. The width of the rectangles increases. It changes by one set of blocks in each step.

 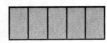

So what comes next? A rectangle with a width of six units would be next.

   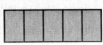 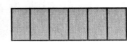

Analyzing Patterns

Look at the pattern. You may be seeing more than you think.

1.	You make an **observation**. What happens when you first look at the set of shapes? You see the details. You see how each shape is alike. And you see what is different.	You might observe: • Rectangles are made of smaller blocks. • The width changes. But the height does not. • The number of blocks in each shape changes. It goes up.
2.	You decide how to make the next shape in the pattern. You do this based on the other shapes.	You figure out that you should put a block at the end of the fourth shape. This will make the next shape. The new shape has six blocks.
3.	You understand the general **rule** for the pattern. You understand what is needed to create as many steps in the pattern as desired.	A possible rule: For each new rectangle, add one block to the previous rectangle.

Working with Geometric Patterns

Sometimes there is more than one rule. Look at the figures in the chart below:

Figure Number	1	2	3	4	5	6
Figure	△	▷	▽	◁		
Outer Portion	4 sides	6 sides	8 sides	10 sides		
Inner Portion	triangle points up	triangle rotates 90° clockwise	triangle rotates 90° clockwise	triangle rotates 90° clockwise		

- Look at the outer portion of each figure. See that it starts with the shape before. Then two sides are added.
- Look at the inner portion. See that the triangles are turned 90° clockwise. It does this with each step.

Sample Rule: Start with the figure before. Draw two more sides to the outside part. Turn the triangle 90° clockwise.

What are the next two shapes?

Here is figure 5. It has 12 sides. The triangle points up.			Here is figure 6. It has 14 sides. The triangle points to the right.

Geometric Patterns in Our Daily Lives

Geometric patterns are everywhere. They are on the floors. They are in art. They help us understand the world around us. The study of fractals is a newer use of geometric patterns. A **fractal** is a shape. It can be split into parts. Each part is a copy of the whole. This is a close-up picture of broccoli. It is a fractal! Understanding fractals can help us understand some patterns in nature.

You Try It

See the figures below. What comes next?

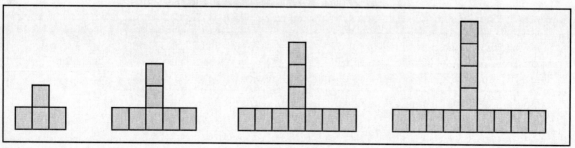

Shaping Up

Look at the cats below. Which cat comes next?

Did you guess a cat with its back arched? "What comes next?" is asked a lot in math. This question can lead to useful thoughts. Sometimes it just leads to a pretty picture.

Basic Facts

A **geometric pattern** uses lines. It uses curves. Or it uses shapes. These must repeat in an orderly way. The figures below make a geometric pattern. The width of the rectangles goes up. It goes up by one set of blocks in each step.

   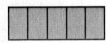

So what comes next? A rectangle with a width of six units would be next.

   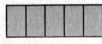 

Analyzing Patterns

Look at the pattern. When you look, you may be seeing more than you think.

1.	You make an **observation**. What happens when you first look at the set of shapes? You notice the details of what you are given. You see what each one has in common. And you see what is different.	You might observe: • Rectangles are made of smaller blocks. • The width changes. But the height does not. • The number of blocks in each rectangle changes.
2.	You decide how to make the next shape in the pattern. You decide this based on the other shapes.	You figure it out by putting a block at the end of the fourth rectangle. This will make the next shape. The new rectangle has six blocks.
3.	You understand the general **rule** for the pattern. You understand what is needed to create as many steps in the pattern as desired.	A possible rule: For each new rectangle, add one block to the previous one.

Working with Geometric Patterns

Sometimes there is more than one rule. Look at the shapes in the chart:

Figure Number	1	2	3	4	5	6
Figure	△	▷	▽	◁		
Outer Portion	4 sides	6 sides	8 sides	10 sides		
Inner Portion	triangle points up	triangle rotates 90° clockwise	triangle rotates 90° clockwise	triangle rotates 90° clockwise		

- Look at the outer portion of each figure. Notice that two sides are added to the previous figure.
- Look at the inner portion. Notice that the triangles are rotated 90° clockwise with each step.

Sample Rule: Start with the previous figure. Draw two more sides to the outside portion. Then, rotate the triangle 90° clockwise.

What are the next two figures?

Here is figure 5. It has 12 sides. The triangle points up.			Here is figure 6. It has 14 sides. The triangle points to the right.

Geometric Patterns in Our Daily Lives

From floors, to art, to understanding the world around us—geometric patterns are everywhere. Among the newer applications is the study and use of fractals. A **fractal** is a shape that can be split into parts, each of which is a copy of the whole. This is a close-up picture of broccoli. It is a fractal! Understanding fractals can help us understand some patterns in nature.

You Try It

Look at the figures below. What comes next?

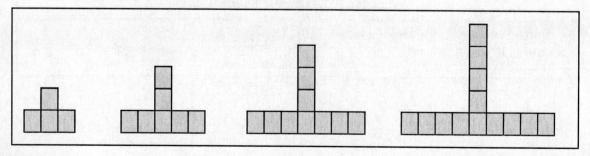

Shaping Up

Look at the cats below. Which cat comes next?

Did you guess a cat arching its back should be next? "What comes next?" is a question often asked in mathematics, and can lead to some valuable insights. Sometimes the question simply leads to an image that is enjoyable to look at.

Basic Facts

A **geometric pattern** uses lines, curves, or shapes. These must be repeated in some sort of orderly fashion. The figures below form a geometric pattern. Notice the width of the rectangles increases, or goes up, by one set of blocks in each subsequent step.

   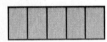

Can you determine what comes next? A rectangle with a width of six units follows.

   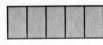 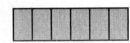

Analyzing Patterns

In studying the pattern, you probably evaluate much more than you are aware of.

1.	You make an **observation**. When you first look at the set of rectangles, you notice the details of what you are given. You notice what each figure has in common and what is different about each figure.	Your observation might include: • Rectangles are made of smaller blocks. • The width changes but the height does not. • The number of blocks in each rectangle changes.
2.	You decide how to create the next item in the pattern. You decide this based on the previous figures.	You figure out that if you put a block at the end of the fourth rectangle, you have the next rectangle. The new rectangle has six blocks.
3.	You understand the general **rule** for the pattern. You understand what is needed to create as many steps in the pattern as desired.	A possible rule: For each new rectangle, add one block to the previous rectangle.

Working with Geometric Patterns

Sometimes there is more than one rule at work. Observe the figures in the chart:

Figure Number	1	2	3	4	5	6
Figure	△	▷	▽	◁		
Outer Portion	4 sides	6 sides	8 sides	10 sides		
Inner Portion	triangle points up	triangle rotates 90° clockwise	triangle rotates 90° clockwise	triangle rotates 90° clockwise		

- Look at the outer portion of each figure. Notice that two sides are added to the previous figure.
- Look at the inner portion. Notice that the triangles are rotated 90° clockwise with each step.

Sample Rule: Start with the previous figure. Draw two more sides to the outside portion. Then, rotate the triangle 90° clockwise.

What are the next two figures?

Here is figure 5. It has 12 sides. The triangle points up.			Here is figure 6. It has 14 sides. The triangle points to the right.

Geometric Patterns in Our Daily Lives

From floor patterns, to art, to envisioning the world around us, geometric patterns abound. A newer application of geometry is the study of fractals. A **fractal** is a shape that can be split into parts, each of which is a replica of the whole. This is a close-up picture of broccoli. It is a fractal! Understanding fractals can help us dissect intricate patterns in nature.

You Try It

Investigate the figures below to determine what should come next.

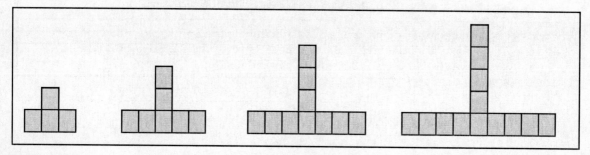

Shaping Up

Examine the cats below and determine which image comes next in the sequence.

Did you correctly guess a cat with its back arched would be next? "What comes next?" is a question often asked in mathematics because it leads to valuable insights. Or, the question simply leads to an interesting image that we enjoy looking at.

Basic Facts

A **geometric pattern** uses lines, curves, or shapes that are repeated in an orderly and predictable way. The figures below form a geometric pattern. Notice the ultimate width of the rectangles increase by one set of blocks in each iteration of the pattern.

So what is the next consecutive figure? A rectangle with a width of six units follows.

   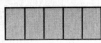 

Analyzing Patterns

In looking at the pattern, you probably do much more than you realize.

1.	Immediately, you make an **observation**. When you first discern the set of rectangles, you zero in on the details. You surmise each figure's commonalities and also the differences between each one.	Your observation might include: • Rectangles are made of smaller blocks. • The width of each figure changes but the height does not. • The number of blocks in each rectangle increases.
2.	You deduce how to create the next item in the pattern based on your perception of the previous figures.	If you add one block at the end of the fourth rectangle, you will create the subsequent rectangle in the series. The new rectangle consists of six blocks.
3.	You see the general **rule** for the pattern. You understand what is needed to create as many iterations of the pattern as desired.	A possible rule: For each new rectangle, add one block to the previous rectangle.

Working with Geometric Patterns

Sometimes more than one rule is applicable. Look at the figures in the chart

Figure Number	1	2	3	4	5	6
Figure	△	▷	▽	◁		
Outer Portion	4 sides	6 sides	8 sides	10 sides		
Inner Portion	triangle points up	triangle rotates 90° clockwise	triangle rotates 90° clockwise	triangle rotates 90° clockwise		

- Investigate the outer portion of each figure: two sides are added to the previous figure.
- Investigate the inner portion: the triangles are rotated 90° clockwise with each progressive step.

Sample Rule: Start with the previous figure. Draw two more sides to the outside portion. Then, rotate the triangle 90° clockwise.

What are the next two figures?

Here is figure 5. It has 12 sides. The triangle points up.			Here is figure 6. It has 14 sides. The triangle points to the right.

Geometric Patterns in Our Daily Lives

From floors, to art, to conceptualizing the world around us, geometric patterns are everywhere. Among the newer applications is the study and use of fractals. A **fractal** is a shape that can be split into parts, each of which is a copy of the whole. This is a magnified photograph of broccoli. It is a fractal! Understanding fractals can help us understand some patterns in nature.

You Try It

Consider the figures below and answer the question, "What comes next?"

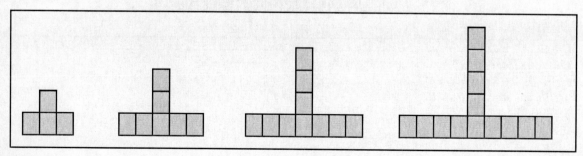

Sometimes the Change Is Consistent

One dragonfly has 6 legs. Think of two dragonflies on a leaf. There are 12 legs. What if there are three? There are 18 legs. How many legs would there be with four? What pattern do you see?

Basic Facts

Think of the pattern 2, 4, 6, 8, 10. This list makes up a **sequence**. That is a list that follows a predictable rule. This is also called a **numerical pattern**. It lets you use one number to find the next number. Here is a list of even numbers. We must add 2 to each number. This will get the next number.

$$\overset{+2}{\frown}\overset{+2}{\frown}\overset{+2}{\frown}\overset{+2}{\frown}\overset{+2}{\frown}$$
2, 4, 6, 8, 10, 12,

A **rule** shows how to move from one number to the next. Here is one way to say the rule: "To find the next number, add 2 to the number before it."

It would help to have a way to say, "The position of a number within a numerical pattern." There is a way. We use the word *term*. And we use *term number* to say this.

- Each term is a number in the pattern. Each number in the pattern 1, 3, 5, 7, 9, 11, 13 is a term. *1* is a term. *3* is a term. *9* is a term. *11* is a term, too.
- The term number is where each term belongs in the pattern. It tells the order. *1* is the first term. *3* is the second term. *5* is the third term. *9* is the fifth term. What would the eighth term be? We need to find 13 + 2 for the answer. We see that 15 would be the eighth term.

Look at the next chart. What is the rule? What is the sixth term? What is the seventh term?

Term Number	1	2	3	4	5	6	7
Term	88	77	66	55	44		

Rule: Subtract 11 from the term before.

sixth term: 33   seventh term: 22

Looking Further

Numerical patterns can start with any number. Look at this pattern:

5, 8, 11, 14, 17,…

What do you see? Do you see that 8 − 5 = 3? This is the second term minus the first term.

So, 5 + 3 = 8, and 8 + 3 = 11
Keep checking: 11 + 3 = 14. 14 + 3 = 17.
The sixth term would be 20. This is because 17 + 3 = 20
The rule is: To find the next term, add 3 to the term before it.

Any operation can be used in a numerical pattern. To find the next term:

99, 97, 95, 93…	Subtract 2 from the term before
2, 6, 18, 54…	Multiply the term before by 3
10,000, 1,000, 100, 10…	Divide the term before by 10

Figuring It Out

Look at the pattern 4, 8, 16, ____, 64, ____, 256, 512, ____,….

What numbers belong in each blank? How can you tell?

1.	You make an **observation**. Try out different rules. See if they work.	First, try 8 + 4 = 12 (Our sequence is not 4, 8, 12…) Then try 8 × 2 = 16 (Our sequence is 4, 8, 16…)
2.	When you think you have the **rule**, check it. Try it everywhere you can.	Multiplying the seventh term by 2 gives the eighth term. 256 × 2 = 512
3.	Understand the rule.	For the next term, multiply the previous term by 2.
4.	Use the rule.	×2 ×2 ×2 ×2 ×2 ×2 ×2 ×2 4, 8, 16, 32, 64, 128, 256, 512, 1,024, …

Numerical Patterns in Our Daily Lives

This is a bacterial cell. It is known as bubonic plague. This one cell becomes two cells. It splits. Those two cells grow. They grow into a full-size cell. Then, each of those cells split. Now there are four cells. Four become eight. Eight become 16. And so on. They follow a numerical pattern. It is 1, 2, 4, 8, 16, 32…. What rule does it follow? The rule is "find the next term by multiplying the term before it by 2."

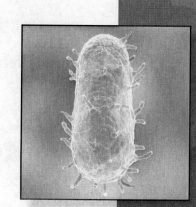

You Try It

Look at this numerical pattern: 128, 64, 32, 16, ___, ___, ___ . Find the rule. Then fill in the blanks. Be sure to check for each operation (+, −, ×, ÷).

Sometimes the Change Is Consistent

One dragonfly has 6 legs. Think of two dragonflies on a leaf. There are 12 legs. What if there are three on a leaf? There are 18 legs. How many legs would there be with four dragonflies? What pattern do you see?

Basic Facts

Think of the pattern 2, 4, 6, 8, 10. This list makes up a **sequence**. This is a list that follows a predictable rule. Another term for it is a **numerical pattern**. It lets you use one number to find the next number. Here is a list of even numbers. Note that 2 must be added to each number in the list to get the next number.

$$\overset{+2}{\frown}\ \overset{+2}{\frown}\ \overset{+2}{\frown}\ \overset{+2}{\frown}\ \overset{+2}{\frown}$$
$$2,\ 4,\ 6,\ 8,\ 10,\ 12,$$

A **rule** must be followed to move from one number to the next. Here is one way to say the rule: "To find the next number, add 2 to the number before it."

It would help to have a way to say, "The position of a number in a numerical pattern." There is a very useful way. We use the words *term* and *term number* to say this.

- Each term is a number that is in the pattern. So each number in the pattern 1, 3, 5, 7, 9, 11, 13 is a term. *1* is a term. *3* is a term. And so on.

- The term number is where each term belongs in the pattern. The *1* is the first term. *3* is the second term. *5* is the third term. What would the eighth term be? We need to find 13 + 2 for the answer. And we see that 15 would be the eighth term.

Look at the next chart. What is the rule? What is the sixth term? What is the seventh term?

Term Number	1	2	3	4	5	6	7
Term	88	77	66	55	44		

Rule: Subtract 11 from the previous term.

sixth term: 33   seventh term: 22

Looking Further

Numerical patterns can start with any number. Look at this pattern:

5, 8, 11, 14, 17,…

Notice that 8 − 5 = 3 (second term − first term).
So, 5 + 3 = 8, and 8 + 3 = 11.
Keep checking: 11 + 3 = 14. 14 + 3 = 17.
The sixth term would be 20 because 17 + 3 = 20.
The rule is: To find the next term, add 3 to the term before it.

Any operation can be used in a numerical pattern. To find the next term:

99, 97, 95, 93…	Subtract 2 from the term before
2, 6, 18, 54…	Multiply the term before by 3
10,000, 1,000, 100, 10…	Divide the term before by 10

Figuring It Out

Look at the pattern 4, 8, 16, ____, 64, ____, 256, 512, ____,…

What numbers belong in each blank? How can you tell?

1.	You make an **observation**.	First, try 8 + 4 = 12 (Our sequence is not 4, 8, 12…) Then try 8 × 2 = 16 (Our sequence is 4, 8, 16…)
2.	When you think you have the **rule**, check it everywhere possible.	Multiplying the seventh term by 2 gives the eighth term. 256 × 2 = 512
3.	Understand the rule.	For the next term, multiply the previous term by 2.
4.	Use the rule.	×2 ×2 ×2 ×2 ×2 ×2 ×2 ×2 4, 8, 16, 32, 64, 128, 256, 512, 1024,…

Numerical Patterns in Our Daily Lives

This bacterial cell is known as bubonic plague. This single cell becomes two cells by splitting. Those two cells then grow into full-sized cells. Then, each of those two cells split. That makes a total of four cells. Four become eight. Eight become sixteen. And so it goes. The basic growth may be shown by the numerical pattern 1, 2, 4, 8, 16, 32…. This pattern follows the rule: "find the next term by multiplying the term before it by 2."

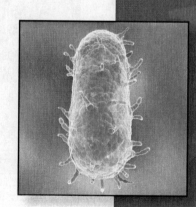

You Try It

Look at this numerical pattern: 128, 64, 32, 16, ___, ___, ___ . Find the rule. Then fill in the blanks. Be sure to check for each operation (+, −, ×, ÷).

Sometimes the Change Is Consistent

One dragonfly has 6 legs. If two dragonflies are perched on a leaf, there are 12 legs on the leaf. If three dragonflies congregate, there are 18 legs. How many legs would there be with four dragonflies? What pattern is emerging in the number of legs on the leaf?

Basic Facts

Think of the pattern 2, 4, 6, 8, 10. This is a **sequence**, or a list that follows a predictable rule. It is also called a **numerical pattern**. A numerical pattern follows a predictable rule, and one number contributes to generating the next number. Notice in this list, 2 must be added to each number to generate the subsequent number.

2, 4, 6, 8, 10, 12,

Notice that a **rule** must be established and applied consistently to continue the pattern. One way to delineate the rule is: "To determine the next number, add 2 to the previous number."

A convenient way to say "the position of a number within a numerical pattern" is to use the words *term* and *term number* to make this distinction.

- Each term is a number that helps to make up the numeric pattern. So each number in the pattern 1, 3, 5, 7, 9, 11, 13 is a term.
- The term number identifies where each term belongs in the numeric pattern. So, *1* is the first term, *3* is the second term, *5* is the third term. If this pattern continued, what would the eighth term be? With 13 + 2, we can find that 15 would be the eighth term.

Look at the next chart. Can you determine the rule? What is the sixth term? What is the seventh term?

Term Number	1	2	3	4	5	6	7
Term	88	77	66	55	44		

Rule: Subtract 11 from the previous term.

sixth term: 33   seventh term: 22

Looking Further

Numerical patterns can start with any number. Look at this pattern:

$$5, 8, 11, 14, 17,\ldots$$

Notice that 8 − 5 = 3 (second term − first term).
    So, 5 + 3 = 8, and 8 + 3 = 11.
Keep checking: 11 + 3 = 14. 14 + 3 = 17.
    The sixth term would be 20 because 17 + 3 = 20.
The rule is: To find the next term, add 3 to the previous term.

Any operation can be utilized in a numerical pattern. Determine the next term:

99, 97, 95, 93…	Subtract 2 from the term before
2, 6, 18, 54…	Multiply the term before by 3
10,000, 1,000, 100, 10…	Divide the term before by 10

Figuring it Out

Look at the pattern 4, 8, 16, ____, 64, ____, 256, 512, ____,….

What numbers belong in each blank? How can you verify your prediction?

1.	You make an **observation**.	First, try 8 + 4 = 12 (Our sequence is not 4, 8, 12…) Then try 8 × 2 = 16 (Our sequence is 4, 8, 16…)
2.	When you think you have the **rule**, check it everywhere possible.	Multiplying the seventh term by 2 gives the eighth term. 256 × 2 = 512
3.	Understand the rule.	For the next term, multiply the previous term by 2.
4.	Use the rule.	×2 ×2 ×2 ×2 ×2 ×2 ×2 ×2 4, 8, 16, 32, 64, 128, 256, 512, 1,024, …

Numerical Patterns in Our Daily Lives

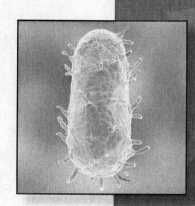

This bacterial cell is known as bubonic plague. This single cell becomes two cells by splitting. Those two cells then grow into full-sized cells. Then, each of those two cells split. That makes a total of four cells. Four become eight. Eight become sixteen. And so it goes. The basic growth may be shown by the numerical pattern 1, 2, 4, 8, 16, 32…. This pattern follows the rule: "find the next term by multiplying the term before it by 2."

You Try It

Look at this numerical pattern: 128, 64, 32, 16, ___, ___, ___ . Establish and apply the rule. Then fill in the blanks with the appropriate numbers. Be sure to verify the possible application of each operation (+, −, ×, ÷).

Sometimes the Change Is Consistent

It's established that one dragonfly has 6 legs; thus, if two dragonflies perch on a leaf, there are 12 legs. If three dragonflies congregate, there are 18 legs. Determine how many legs there would be with four dragonflies. Can you establish a pattern to classify the relationship between legs and dragonflies?

Basic Facts

This pattern 2, 4, 6, 8, 10 constitutes a **sequence**, which describes a list that follows a predictable rule. This can also be referred to as a **numerical pattern** because it also requires that one number generates the next number in the sequence. Notice that in this list of even numbers, 2 must be added to each number in the list to generate the subsequent number.

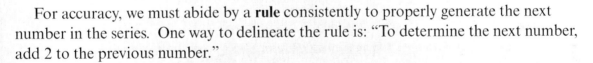

For accuracy, we must abide by a **rule** consistently to properly generate the next number in the series. One way to delineate the rule is: "To determine the next number, add 2 to the previous number."

It would help to have an expression that stands for: "The position of a number within a numerical pattern." There is an expression we use. We use the words *term* and *term number* to make this distinction.

- Each term is a number that is essential to the numeric pattern. So each number in the pattern 1, 3, 5, 7, 9, 11, 13 is a term.
- The term number establishes the location of each term within the numeric pattern. So, *1* is the first term, *3* is the second term, *5* is the third term. If this pattern continued, what would the eighth term be? With 13 + 2, we can find that 15 would be the eighth term.

Evaluate the chart to determine the rule. Verify your prediction by solving for the sixth term. Can you determine the veracity of the pattern and find the seventh term?

Term Number	1	2	3	4	5	6	7
Term	88	77	66	55	44		

Rule: Subtract 11 from the term before.

sixth term: 33   seventh term: 22

Looking Further

Numerical patterns can start with any number. Look at this pattern:

$$5, 8, 11, 14, 17,\ldots$$

Notice that 8 − 5 = 3 (second term − first term).
 So, 5 + 3 = 8, and 8 + 3 = 11.
Keep checking: 11 + 3 = 14. 14 + 3 = 17.
 The sixth term would be 20 because 17 + 3 = 20.
The rule is: To determine the next term, add 3 to the previous term in the series.

Any mathematical operation can be used in a numerical pattern. For example, in the following patterns, to determine the next term:

99, 97, 95, 93…	Subtract 2 from the term before
2, 6, 18, 54…	Multiply the term before by 3
10,000, 1,000, 100, 10…	Divide the term before by 10

Figuring it Out

Look at the pattern 4, 8, 16, ____, 64, ____, 256, 512, ____,…

What numbers belong in each blank? How can you tell?

1.	You make an **observation**. Compare terms and try different numbers.	First, try 8 + 4 = 12 (Our sequence is not 4, 8, 12…) Then try 8 × 2 = 16 (Our sequence is 4, 8, 16…)
2.	Determine a **rule**. When you think you have it, check it everywhere possible.	Multiplying the seventh term by 2 gives the eighth term. 256 × 2 = 512
3.	Understand the rule that you have found.	For the next term, multiply the previous term by 2.
4.	Use the rule.	×2 ×2 ×2 ×2 ×2 ×2 ×2 ×2 4, 8, 16, 32, 64, 128, 256, 512, 1,024, …

Numerical Patterns in Our Daily Lives

This bacterial cell is known as bubonic plague. This single cell duplicates itself into two identical cells by splitting. Those two cells mature into full-sized cells. Then, each of those two cells split. That makes a total of four cells. Four become eight. Eight become sixteen. And so it goes. The exponential growth may be shown by the numerical pattern 1, 2, 4, 8, 16, 32…. This pattern follows the rule: "find the next term by multiplying the term before it by 2."

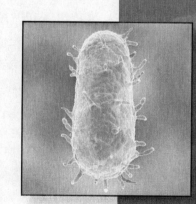

You Try It

Evaluate this numerical pattern: 128, 64, 32, 16, ___, ___, ___ . Establish the parameters of the rule, then apply it to fill in the blanks. Be sure to verify the use of each possible operation (+, −, ×, ÷).

Sometimes the Change Changes

Look at the figures below. What should come next?

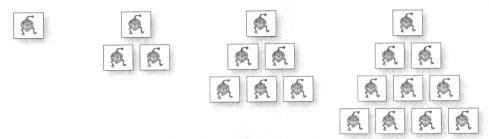

Basic Facts

These figures form a **geometric pattern**. We can look at the figures. Then, we can predict what comes next. What did you find that the fifth figure would be? It is a triangle with the bottom row having five monsters.

The first figure has one shape. The second has three shapes. The third has six shapes. And the fourth has 10 shapes. We can show the total number of monsters in each group. Here is the series: 1, 3, 6, 10,.... This is a list. It follows a predictable rule. So, it is a **sequence**. This is a list of numbers. And the numbers follow a predictable rule. So, it is a **numerical pattern**.

Back to 1, 3, 6, 10…

So what about this sequence? It goes up. But the amount changes from term to term.

$$1,\ 3,\ 6,\ 10,\ \_\_\_,\ \_\_\_,\ \_\_\_,\ \_\_\_,$$
$$+2\ +3\ +4$$

Look at the amount of change. In this sequence, the amount from one term to the next starts at 2. Then it goes to 3. And then it is 4. This amount of change is a pattern! We can see that we add 2. Then we add 3. Then we add 4. Then we add 5. Then we add 6. And then we add 7. Then we add 8. And it keeps going.

$$+2\ +3\ +4\ +5\ +6\ +7\ +8\ +9$$
$$1,\ 3,\ 6,\ 10,\ 15,\ 21,\ 28,\ 36,\ 45$$

How to Figure Out a Rule

Ask a question. "What do I add, subtract, multiply, or divide to move from one term to the next?" Look at that answer. You will see the pattern.

Sometimes there are patterns that are not easy to see. There are many things that can make it hard. Start by checking to see if there is a pattern of change from one number to the next. This can help.

Working Through Examples

You have worked with patterns in the past. So, you should follow these steps:

1. Find your possible pattern.
2. When you think you have the rule, check it everywhere possible.
3. Understand the rule.
4. Use the rule.

Look at this example. 3, 5, 11, 21, 35, 53, ____, ____, ____,…

What are the next three numbers?

These numbers are going up. As you look at the change from one number to the next, make note of that change.

$$3,\ 5,\ 11,\ 21,\ 35,\ 53,\ \_\_,\ \_\_,\ \_\_,\ldots$$
$$+2\ +6\ +10\ +14\ +18$$

Look at the changes you noted. The numbers went up by 2, 6, 10, 14, 18…. That is a pattern!

The differences between terms went up by 4.

So what comes after 53? Look at the last term. The difference was 18.

So, for the next term, the difference will be $18 + 4 = 22$. We can use that difference to find the 7th term.

$$3,\ 5,\ 11,\ 21,\ 35,\ 53,\ \underline{75},\ \underline{101},\ \underline{131}$$
$$+2\ +6\ +10\ +14\ +18\ +22\ +26\ +30$$

Numerical Patterns in Our Daily Lives

Imagine a toy car. It starts 2 feet from its controller. It is speeding up. There is a pattern of 2, 5, 11, 20, 32, 47, 65, 86,…. This shows how many feet the car is from the remote controller at each second. We can use the pattern. It shows how to find how far the car went. We can tell how far it was after 10 seconds. We can see how far it was after 15 seconds. We can use the pattern for any time.

Being able to see patterns is important. People work with patterns all the time. They are used in mathematics. They are used in science. They are used in business. And they are used in many other areas.

You Try It

Look at the pattern from the toy car above. Find the next three terms. Describe how you figured out the rule.

Sometimes the Change Changes

Look at the figures below. What should come next?

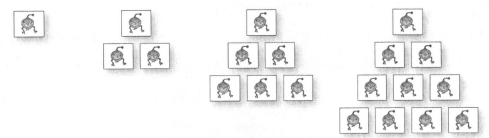

Basic Facts

The figures above form a **geometric pattern**. We can look at the figures. Then, we can predict what comes next. Did you find that the fifth figure would be a triangle with the bottom row having five monsters?

The first figure has one shape. The second has three shapes. The third has six shapes. And the fourth has 10 shapes. We can show the total number of monsters in each figure as 1, 3, 6, 10,…. This is a list that follows a predictable rule. So, it is a **sequence**. Since this is a list of numbers that follow a predictable rule, it is a **numerical pattern**.

Back to 1, 3, 6, 10…

So what about this sequence? The amount that it goes up changes from term to term.

$$1,\ 3,\ 6,\ 10,\ \_\_\_,\ \_\_\_,\ \_\_\_,\ \_\_\_,$$
$$+2\ +3\ +4$$

Look at the amount of change. In this sequence, the amount from one term to the next starts at 2. Then it goes to 3. And then it is 4. This amount of change is a pattern! We can see that we add 2. Then we add 3. Then we add 4, then 5, then 6, then 7, then 8, and so on.

$$+2\ +3\ +4\ +5\ +6\ +7\ +8\ +9$$
$$1,\ 3,\ 6,\ 10,\ 15,\ 21,\ 28,\ 36,\ 45$$

How to Figure Out a Rule

Ask yourself, "What do I add, subtract, multiply, or divide to move from one term to the next?" Look for a pattern in that answer.

Sometimes there are patterns that are not easy to see. There are many possibilities. But it helps to start by checking to see if there is a pattern of change from one number to the next.

Working Through Examples

You have worked with patterns in the past. So, you should follow these steps:

1. Find your possible pattern.
2. When you think you have the rule, check it everywhere possible.
3. Understand the rule.
4. Use the rule.

Look at this example. 3, 5, 11, 21, 35, 53, ____, ____, ____,…

What are the next three numbers?

Notice that these numbers are going up. As you look at the change from one number to the next, make note of that change.

$$3,\ 5,\ 11,\ 21,\ 35,\ 53,\ \_\_\_,\ \_\_\_,\ \_\_\_,\ldots$$
$$+2\ +6\ +10\ +14\ +18$$

Look at the changes you noted. The numbers went up by 2, 6, 10, 14, 18…. That is a pattern!

The differences between terms went up by 4.

So what comes after 53? Look at the last term. The difference was 18.

So, for the next term, the difference will be 18 + 4 = 22. We can use that difference to find the 7th term.

$$3,\ 5,\ 11,\ 21,\ 35,\ 53,\ \underline{75},\ \underline{101},\ \underline{131}$$
$$+2\ +6\ +10\ +14\ +18\ +22\ +26\ +30$$

Numerical Patterns in Our Daily Lives

Imagine a toy car starting 2 feet from its remote control operator. It is speeding up. The pattern of 2, 5, 11, 20, 32, 47, 65, 86… shows how far, in feet, the car is from the remote controller at each second. We can use the pattern to find how far the car traveled. We can tell how far it was after 10 seconds. We can see how far it was after 15 seconds, and so on.

The ability to recognize and work with patterns is important. People work with patterns in mathematics, science, business, and many other areas.

You Try It

Find the next three terms in the pattern from the toy car above. Describe how you figured out the rule.

Sometimes the Change Changes

Look at the figures below. What is the next figure in the pattern?

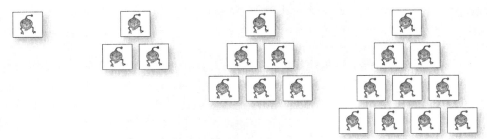

Basic Facts

The figures above form a **geometric pattern**. By looking at the given figures, we can find what comes next. Did you find that the fifth figure would be a triangle with the bottom row having five shapes?

Notice that the first figure has one shape, the second figure has three shapes, the third figure has six shapes, and the fourth figure has 10 shapes. We can show the total number of monsters in each figure as 1, 3, 6, 10,…. This is a list that follows a predictable rule so it is a **sequence**. Since this is a list of numbers that follow a predictable rule, it is a **numerical pattern**.

Back to 1, 3, 6, 10…

So what about this sequence? The amount of increase changes from term to term.

$$1,\ 3,\ 6,\ 10,\ \_\_\_,\ \_\_\_,\ \_\_\_,\ \_\_\_,$$
$$+2\ +3\ +4$$

Look at the amount of change. In this sequence the amount from one term to the next is 2, then 3, and then 4. This amount of change is a pattern in itself! We can see that we add 2, then 3, then 4, then 5, then 6, then 7, then 8, and so on.

$$+2\ +3\ +4\ +5\ +6\ +7\ +8\ +9$$
$$1,\ 3,\ 6,\ 10,\ 15,\ 21,\ 28,\ 36,\ 45$$

How to Figure Out a Rule

To figure out a pattern's rule, ask yourself, "What do I add, subtract, multiply, or divide to move from one term to the next?" Look for a pattern in that answer.

Sometimes there are patterns that are not immediately straightforward. There are many possibilities. But it certainly helps to start by checking to see if there is a pattern of some sort of change from one number to the next.

Working Through Examples

Having worked with patterns in the past, you know you should follow these steps:

1. Determine your possible pattern.
2. When you think you have the rule, check it everywhere possible.
3. Understand the rule.
4. Use the rule.

Consider this example: 3, 5, 11, 21, 35, 53, ____, ____, ____,…

What are the next three numbers in this sequence?

Notice that these numbers are increasing (going up). As you look at the change from one number to the next, make note of that change.

$$3, \underset{+2}{\vee} 5, \underset{+6}{\vee} 11, \underset{+10}{\vee} 21, \underset{+14}{\vee} 35, \underset{+18}{\vee} 53, \_\_\_, \_\_\_, \_\_\_, \ldots$$

Look at the changes you noted. The numbers increased by 2, 6, 10, 14, 18…. That's a pattern!

The differences between terms increased by 4.

So what comes after 53? For the last term, the difference was 18.

So, for the next term, the difference will be 18 + 4 = 22. We can use that difference to find the seventh term.

$$3, \underset{+2}{\vee} 5, \underset{+6}{\vee} 11, \underset{+10}{\vee} 21, \underset{+14}{\vee} 35, \underset{+18}{\vee} 53, \underset{+22}{\vee} \underline{75}, \underset{+26}{\vee} \underline{101}, \underset{+30}{\vee} \underline{131}$$

Numerical Patterns in Our Daily Lives

Imagine a toy car starting 2 feet from its remote control operator. It is speeding up. The pattern of 2, 5, 11, 20, 32, 47, 65, 86… shows how far, in feet, the car is from the operator at each second. We can use the pattern to find how far the car traveled after 10 seconds, 15 seconds, and so on.

The ability to see and work with patterns is important in mathematics, science, business, and many other areas.

You Try It

What are the next three terms in the pattern from the toy car above? Describe how you figured out the rule.

Sometimes the Change Changes

Examine the figures below and determine the next figures in the pattern.

Basic Facts

The figures above make a **geometric pattern**. By looking at the given figures, we can predict what comes next. Did you determine that the fifth figure would be a triangle with five shapes in the bottom row?

Notice that the first figure has one shape, the second figure has three shapes, the third figure has six shapes, and the fourth figure has 10 shapes. We can express the total number of monsters in each figure as 1, 3, 6, 10,…. This list follows a predictable rule so it is a **sequence**. Additionally, this predictable rule is a **numerical pattern**.

Back to 1, 3, 6, 10…

So what can we notice about this sequence? If you look carefully, you will see that the amount of increase changes from term to term.

$$1,\ 3,\ 6,\ 10,\ \_\_\_,\ \_\_\_,\ \_\_\_,\ \_\_\_,$$
$$+2\ +3\ +4$$

Consider the amount of change. In this sequence the amount from one term to the next is 2, and then progresses to 3, and then 4. This amount of change is a pattern in itself! We can see that we add 2, then 3, then 4, then 5, then 6, then 7, then 8, and so on to continue the pattern.

$$+2\ +3\ +4\ +5\ +6\ +7\ +8\ +9$$
$$1,\ 3,\ 6,\ 10,\ 15,\ 21,\ 28,\ 36,\ 45$$

How to Figure Out a Rule

To determine a pattern's rule, ask, "What do I add, subtract, multiply, or divide to move from one term to the next?" Search for a pattern within that answer.

Sometimes patterns are not immediately noticeable. Maybe two terms are used to get the next, or perhaps there are two rules that interact in a tricky way. There are many possibilities, so first check to see if there is a pattern of change from number to number.

Working Through Examples

Having worked with patterns in the past, you know you should follow these steps:

1. Determine your possible pattern.
2. When you think you have the rule, check it everywhere possible.
3. Understand the rule.
4. Use the rule.

Consider this example: 3, 5, 11, 21, 35, 53, ____, ____, ____,…

What are the next three numbers in this sequence?

Notice that these numbers are increasing (going up). As you look at the change from one number to the next, make note of that change.

$$3,\ 5,\ 11,\ 21,\ 35,\ 53,\ \_\_,\ \_\_,\ \_\_,\ldots$$
$$+2\ +6\ +10\ +14\ +18$$

Investigate the changes you noted. The numbers increased by 2, 6, 10, 14, 18,…. You've established a pattern!

The differences between terms increased by 4.

So how do you determine what term comes after 53?

For the last term, the difference was 18, so for the next term the difference will be 18 + 4 = 22. We can use that difference to determine the seventh term.

$$3,\ 5,\ 11,\ 21,\ 35,\ 53,\ \underline{75},\ \underline{101},\ \underline{131}$$
$$+2\ +6\ +10\ +14\ +18\ +22\ +26\ +30$$

Numerical Patterns in Our Daily Lives

Imagine a toy car starting 2 feet from its remote control operator, then accelerating. The pattern of 2, 5, 11, 20, 32, 47, 65, 86… describes the distance, in feet, the car travels from the operator at each second. We can use the pattern to find how far the car traveled after 10 seconds, 15 seconds, and so on.

The ability to recognize and work with patterns is imperative in mathematics, science, business, and many other areas.

You Try It

Determine the next three consecutive terms in the pattern from the toy car example above, and describe how you established the rule.

#50716—Leveled Texts for Mathematics: Algebra and Algebraic Thinking   © Shell Education

It's All Organized

Amal warms up for practice. He stands at the free-throw line. He tries to shoot 10 baskets. He does this every day. How many tries will he have made after 5 days? Look at the series below. Which number shows this?

10, 20, 30, 40, 50, 60, 70, 80, 90,…

Basic Facts

Think of the example. Each day only has one answer. How many tries had Amal made after day 1? He had made 10 tries. How many after two days? He had tried 20. How many after three? He had tried 30. What about day 5? Could you say that he had tried a total of both 40 and 50 baskets? No. Only one answer is right. How many had he tried after 5 days? He had tried 50.

Remember the word *variable*? A **variable** is a symbol. It is used to stand for a number. How can this help? How many tries had been made after n days? A total of 10 times n baskets have been tried. So n is the variable. We can show this in a table.

Number of Days	1	2	3	4	5	6
Total Number of Baskets	10	20	30	40	50	60

or

Number of Days	Total Number of Baskets
1	10
2	20
3	30
4	40
5	50
6	60

Look at the tables. "Number of Days" shows what we start with. This is the **input**. "Total Number of Baskets" shows the results. This is the **output**. The tables pair up numbers. They take one input and pair it with one output.

Understanding Functions and Function Tables

Function: A pairing of numbers. Each input gives only one output.
Function table: A table where each input gives only one output.

How do you fill in a function table? Use the rule! The middle column shows how it works.

Input	Input + 12	Output
4	4 + 12	16
7	7 + 12	19
13	13 + 12	25
21	21 + 12	33
45	45 + 12	57

Sometimes you have to figure out the rule. Ask: "How do I get from the input to the output?"

Input	?	Output
12		9
15		12
21		18
32		29
56		53

One More Example

Yossi makes yarn animals. She gives them to a hospital for sick kids. She makes the same number each day. Look at the table below. It shows the total she has made. It shows this for any given number of days.

Number of Days	Total Number of Animals
1	x
2	12
6	36
?	72

- Find the rule.
    $12 \div 2 = 6$ or $36 \div 6 = 6$
    Rule: Number of days $\times 6$
- Find the amount of animals for day 1.
    $1 \times 6 = 6$ animals
- Find the number of days for 72 animals.
    $72 \div 6 = 12$

Function Tables in Our Daily Lives

How much water will evaporate in 2 hours? How much in 3? How much in 7 hours? How much in 100? How much money is made by washing 3 cars? How much for 12 cars? How much for 25 cars? Function tables can help us. They give a close-up view of the problem. They are used in business. They are used in medicine. They are used in astronomy, too. They help in many other fields, as well.

You Try It

Number of Games	Cost
1	
2	
4	$32
7	$56
	$88

This function table shows the cost of games. Each game costs the same amount. What is that amount? Fill in the table.

It's All Organized

Each day, Amal warms up for practice. He tries to shoot 10 baskets from the free-throw line. Look at the sequence below. Which number shows his total number of tries after five days?

10, 20, 30, 40, 50, 60, 70, 80, 90,...

Basic Facts

In this example, each day only has one possible answer. How many tries had Amal made after day 1? He had tried 10 baskets. How many after two days? He had tried 20 baskets. How many after three? He had tried 30. Could you say that on day 5 a total of both 40 and 50 baskets had been tried? No. Only one answer is right. After five days, he had tried 50 baskets.

Remember the word *variable*? A **variable** is any symbol used to stand for a number. How can variables help here? After n days, a total of 10 times n baskets have been tried. So n is the variable. We can show our data in a table.

Number of Days	1	2	3	4	5	6
Total Number of Baskets	10	20	30	40	50	60

or

Number of Days	Total Number of Baskets
1	10
2	20
3	30
4	40
5	50
6	60

Look at the tables. "Number of Days" shows what we start with. This is the **input**. "Total Number of Baskets" shows the results. This is the **output**. The tables pair one input with one output.

Understanding Functions and Function Tables

Function: A pairing of numbers in which each input gives only one output.
Function table: A table where each input returns only one output.

How do you fill in a function table? Use the rule! The middle column shows how it works.

Input	Input + 12	Output
4	4 + 12	16
7	7 + 12	19
13	13 + 12	25
21	21 + 12	33
45	45 + 12	57

Sometimes you have to figure out the rule. Ask: "How do I get from the input to the output?"

Input	?	Output
12		9
15		12
21		18
32		29
56		53

One More Example

Yossi makes yarn animals. She gives them to a children's hospital. She makes the same amount each day. The function table below shows the total she has made after any given number of days.

Number of Days	Total Number of Animals
1	x
2	12
6	36
?	72

- Find the rule.
  $12 \div 2 = 6$ or $36 \div 6 = 6$
  Rule: Number of days × 6
- Find the amount of animals for day 1.
  $1 \times 6 = 6$ animals
- Find the number of days for 72 animals.
  $72 \div 6 = 12$

Function Tables in Our Daily Lives

How much liquid will evaporate in 2 hours? How much in 3? How much in 7 hours? How much money is made by washing 3 cars? How much for 12 cars? How much for 25 cars? There are many times when a function table can help us. It gives a close-up view of the problem. Function tables are used in business. They are used in medicine. They are used in astronomy, too. They are used in many other fields.

You Try It

Number of Games	Cost
1	
2	
4	$32
7	$56
	$88

This function table shows the cost of games. Each game costs the same amount. What is that amount per game? Complete the table.

It's All Organized

Each day when Amal is warming up for basketball practice he attempts to shoot 10 baskets from the free-throw line. Which number in the sequence below shows his total number of attempts after five days?

10, 20, 30, 40, 50, 60, 70, 80, 90,…

Basic Facts

In this example, there is only one possible answer for any one day. After day 1, a total of 10 baskets had been attempted. After two days, Amal had tried 20 baskets. After three days, he had tried 30. Could you say that on day 5 a total of both 40 *and* 50 baskets had been attempted? No. Only one answer is correct. After five days, a total of 50 baskets had been attempted.

Remember the term *variable*? A **variable** is any symbol used to represent a number. We can say that after n days, a total of 10 times n baskets have been tried. So n is a variable here. We can show our information in a table.

Number of Days	1	2	3	4	5	6
Total Number of Baskets	10	20	30	40	50	60

or

Number of Days	Total Number of Baskets
1	10
2	20
3	30
4	40
5	50
6	60

In these tables, "Number of Days" shows what we start with. This is the **input**. "Total Number of Baskets" shows the results. This is the **output**. The tables pair each input with an output.

Understanding Functions and Function Tables

Function: A pairing of numbers in which each input can result in only one output.
Function table: A table where each input results in only one output.

How do you fill in a function table? Use the rule! The middle column shows how it works.

Input	Input + 12	Output
4	4 + 12	16
7	7 + 12	19
13	13 + 12	25
21	21 + 12	33
45	45 + 12	57

Sometimes you have to figure out the rule. Ask: "How do I get from the input to the output?"

Input	?	Output
12		9
15		12
21		18
32		29
56		53

One More Example

Yossi makes yarn animals to donate to a children's hospital. She makes the same amount each day. The function table below shows the total she has made after any given number of days.

Number of Days	Total Number of Animals
1	x
2	12
6	36
?	72

- Find the rule.
  $12 \div 2 = 6$ or $36 \div 6 = 6$
  Rule: Number of days × 6
- Find the amount of animals for day 1.
  $1 \times 6 = 6$ animals
- Find the number of days for 72 animals.
  $72 \div 6 = 12$

Function Tables in Our Daily Lives

How much liquid will evaporate in 2, 3, or 7 hours? How much money is made by washing 3, 12, or 25 cars? There are many times when a function table can help us get a close-up view of the problem. Function tables are used in business, medicine, astronomy, and many other fields, as well.

You Try It

Number of Games	Cost
1	
2	
4	$32
7	$56
	$88

This function table shows the cost of games where each game costs the same amount. What is that amount per game? Complete the table.

It's All Organized

Each day when Amal is warming up for basketball practice, he attempts to shoot 10 baskets from the free-throw line. Which number in the sequence below shows his total number of attempts after five consecutive days?

10, 20, 30, 40, 50, 60, 70, 80, 90…

Basic Facts

In this example, there is only one potential answer for any one given day. After day 1, a total of 10 free-throw baskets had been attempted. After two days, Amal had attempted a total of 20 baskets, and after three days he had attempted 30. Could you say that on day 5 a total of both 40 *and* 50 free-throw baskets had been attempted? No. Only one answer is correct. After five days, a total of exactly 50 free-throw baskets had been attempted.

Remember the term *variable*? **A variable** is any symbol used to represent, or stand in for, a number. In this example, we can assign n as a variable and say that "after n days, a total of 10 times n baskets have been attempted." We can also organize and display our information in a table.

Number of Days	1	2	3	4	5	6
Total Number of Baskets	10	20	30	40	50	60

or

Number of Days	Total Number of Baskets
1	10
2	20
3	30
4	40
5	50
6	60

In these tables, "Number of Days" establishes what we start with. We call this the **input**. "Total Number of Baskets" shows the results, which is the **output**. The tables pair up each input with a single output.

Understanding Functions and Function Tables

Function: A pairing of numbers in which each input can result in only one output.
Function table: A table where each input results in only one output.

How do you fill in a function table? Use the rule! The middle column shows how it works.

Input	Input + 12	Output
4	4 + 12	16
7	7 + 12	19
13	13 + 12	25
21	21 + 12	33
45	45 + 12	57

Sometimes you have to figure out the rule. Ask: "How do I get from the input to the output?"

Input	?	Output
12		9
15		12
21		18
32		29
56		53

One More Example

Yossi makes the same number of yarn animals to donate to a children's hospital each day. The function table below shows the total amount she has made after any given number of days.

Number of Days	Total Number of Animals
1	x
2	12
6	36
?	72

- Find the rule.
  $12 \div 2 = 6$ or $36 \div 6 = 6$
  Rule: Number of days × 6
- Find the amount of animals for day 1.
  $1 \times 6 = 6$ animals
- Find the number of days for 72 animals.
  $72 \div 6 = 12$

Function Tables in Our Daily Lives

How much liquid will evaporate in 2, 3, or 7 hours? How much money is made by washing 3, 12, or 25 cars? There are many times when a function table can help us get a close-up view of the problem. Function tables are used in business, medicine, astronomy, and many other fields, as well.

You Try It

Number of Games	Cost
1	
2	
4	$32
7	$56
	$88

This function table displays the cost of games where each game is an equivalent price. Determine the amount per game and complete the table.

#50716—Leveled Texts for Mathematics: Algebra and Algebraic Thinking   © Shell Education

Express It Mathematically

Think of a number. Add 2. Add 5. Subtract 1. Subtract 6. What is your answer? Is it the same as the first number? It should be. Write n in place of your number. Then write your problem this way:

$$n + 2 + 5 - 1 - 6.$$

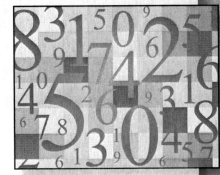

Basic Facts

What is a **mathematical expression**? It is a combination. It has parts. It can have **numbers**. It can have **variables**. And it can have **operations**. Here are some examples:

$$2 + 4 \qquad 2{,}184 - 3 \qquad 7 + 16 \div 4 - 2 \qquad 64 \div 2 - 3$$

A variable is a symbol. It is used to stand for a number. It can be a letter. It can be a shape. It can be any symbol. Here are some:

$$b, \text{ where } b = 2 \qquad \square, \text{ where } \square = 7 \qquad r, \text{ where } r = 4$$

Look at $n + 2 + 5 - 1 - 6$. This is a mathematical expression. It has parts. There are three kind of parts. The variable is n. The operations are $+$ and $-$. The numbers are 2, 5, 1, and 6.

Writing Expressions

Seth is earning money. He wants to go camping. He must pay a $25 fee. Write an expression. Make it show how much money he has left after paying his fee.

Look at the chart. It has expressions. Each shows how much money Seth may earn minus the fee.

$m - \$25$	
$27 − $25	What stays the same?
$35 − $25	*The subtraction of $25 stays the same.*
$42 − $25	What changes?
$67 − $25	*The first number changes.*

Look at the m. It stands for the number that changes. The m can stand for the $27. It can stand for the $35. It can be the $42. And it can stand for the $67. The expression is $m - \$25$. It shows the money Seth will have left.

What Can We Do with this Expression?

Look at $m - \$25$. It shows our problem. m is the money earned. And $25 is the camping fee. Now we have an expression. It is useful. It can be used for any amount of money. It can change.

What Can We Do with this Expression? (cont.)

Question: What is the value of the expression if $m = \$79$?

Work: If $m = \$79$, then $m - \$25 = \$79 - \$25$. Substitute $79 for m.
$\$79 - \$25 = \$54$ Subtract.

Answer: The value is $54.

Understand: What if Seth earns $79? Then he has $54 left after the fee.

Expressions in a Table

Look at each triangle. Each has the three sides. One side length is 50 units long. The second side length is 40 units. The third side length is different for each triangle. Add up all the sides for each triangle. The perimeter is the sum of the three sides.

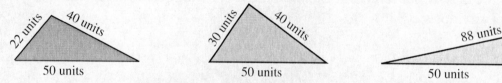

Notice that the third side changes on each triangle. In each case the perimeter equals $40 + 50 +$ the third side. Since the first two sides always add up to 90, the perimeter is always $90 +$ the third side. If we let $s =$ the third side, then $90 + s$ can be used to find the perimeter.

We can make a table. It will show the perimeter for various side lengths. In each case, we substitute the value for s into the expression $90 + s$. We then work the addition to find the perimeter.

Side length (s)	Showing work ($90 + s$)	Perimeter ($90 + s$)
22	90 + 22	112
30	90 + 30	120
88	90 + 88	178

Side length (s)	22	30	88
Perimeter ($90 + s$)	112	120	178

Expressions in Our Daily Lives

Expressions are used in just about any job you can imagine. They show how numbers, variables, and operations work together. How much paint is needed? How much money will be made? How much rocket fuel should be used? Expressions help us understand all these situations.

You Try It

Karen is decorating cookies for a party. She will set 12 cookies aside for a neighbor. Write an expression to show how many cookies she has left for the party. Let $c =$ the number of cookies she makes.

Express It Mathematically

Think of a number. Add 2. Add 5. Subtract 1. Subtract 6. Is your answer the same as the number you started with? It should be. Let us write *n* in place of your number. Then your problem can be written this way:

$$n + 2 + 5 - 1 - 6.$$

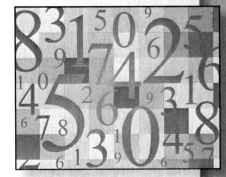

Basic Facts

What is a **mathematical expression**? It is any combination of **numbers**, **variables**, and **operations**. Here are some examples:

$$2 + 4 \qquad 2{,}184 - 3 \qquad 7 + 16 \div 4 - 2 \qquad 64 \div 2 - 3$$

A variable is a symbol used to stand for a number. It can be a letter. It can be a shape. It can be any symbol. Here are some:

$$b, \text{ where } b = 2 \qquad \Box, \text{ where } \Box = 7 \qquad r, \text{ where } r = 4$$

Look again at $n + 2 + 5 - 1 - 6$. This is a mathematical expression. The variable is n. The operations are $+$ and $-$. The numbers are 2, 5, 1, and 6.

Writing Expressions

Seth is earning money. He wants to go camping. He must pay a $25 fee. Write an expression. Make it show the amount of money he has left after paying his fee.

Look at these expressions. Each shows Seth's possible earned money minus the fee.

$m - \$25$	
$27 - $25	What stays the same?
$35 - $25	*The subtraction of $25 stays the same.*
$42 - $25	What changes?
$67 - $25	*The first number changes.*

Look at the *m*. It stands for the number that changes. The *m* can stand for the $27. It can stand for the $35. It can be the $42. And it can stand for the $67. Look at this expression. $m - 25. It shows the money Seth will have left after paying his fee.

What Can We Do with this Expression?

We know the expression $m - \$25$ shows our problem. *m* equals the money earned. And $25 is the camping fee. We now have an expression. It is useful. It can be used for any amount of money.

What Can We Do with this Expression? (cont.)

Question: What is the value of the expression if $m = 79$?

Work: If $m = \$79$, then $m - \$25 = \$79 - \$25$.   Substitute $79 for m.
$\$79 - \$25 = \$54$ Subtract.

Answer: The value is $54.

Understand: What if Seth earns $79? Then he has $54 left after the fee.

Expressions in a Table

Look at these shapes. Each has three sides. One side is 50 units long. Each has a second side. Its length is 40 units. The third side is different. It changes for each shape. Add up all 3 sides for each triangle. That is the perimeter.

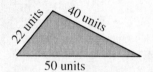
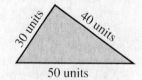
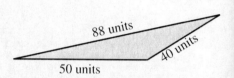

The third side changes on each triangle. So, the perimeter is $40 + 50 +$ the third side. The first two sides are always the same. They add up to 90. So it can be written $90 +$ the third side. Let $s =$ the third side. So, $90 + s$ will be the perimeter.

We can make a table. It will show different side lengths. We assign the value for s into the expression $90 + s$. Then we add. This determines the perimeter.

Side length (s)	Showing work ($90 + s$)	Perimeter ($90 + s$)
22	90 + 22	112
30	90 + 30	120
88	90 + 88	178

Side length (s)	22	30	88
Perimeter ($90 + s$)	112	120	178

Expressions in Our Daily Lives

Expressions are used everywhere. They are involved in just about any job you can imagine. They show how numbers, variables, and operations work together. How much paint is needed? How much money will be made? How much rocket fuel should be used? Expressions help us understand all of these.

You Try It

Karen is going to a party. She is making cookies. She will set 12 cookies aside. These are for a neighbor. Write an expression. Make it show how many cookies she has left for the party. Let $c =$ the number of cookies she makes.

Express It Mathematically

Think of a number. Add 2. Add 5. Subtract 1. Subtract 6. Is your result the same as the number you started with? It should be. If we write n in place of your number, your calculations can be written as:

$$n + 2 + 5 - 1 - 6.$$

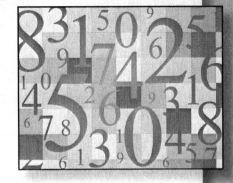

Basic Facts

A **mathematical expression** is any combination of **numbers**, **variables**, and **operations**. Here are some examples:

$$2 + 4 \qquad 2{,}184 - 3 \qquad 7 + 16 \div 4 - 2 \qquad 64 \div 2 - 3$$

Remember that a variable is any symbol used to represent a number. Here are some examples:

$$b, \text{ where } b = 2 \qquad \square, \text{ where } \square = 7 \qquad r, \text{ where } r = 4$$

Look again at $n + 2 + 5 - 1 - 6$. This is a mathematical expression. The variable is n. The operations are $+$ and $-$. The numbers are 2, 5, 1, and 6.

Writing Expressions

Seth is earning money for a camping trip. He must pay a $25 fee for the location. Write an expression showing the amount of money he has left after paying his fee.

Look at these expressions. Each shows Seth's possible earned money minus the fee.

$m - \$25$	
$27 – $25	What stays the same?
$35 – $25	*The subtraction of $25 stays the same.*
$42 – $25	What changes?
$67 – $25	*The first number changes.*

The variable m represents the number that changes. The letter m can be written in place of the $27, the $35, the $42, and the $67. The expression $m - \$25$ shows the money Seth will have left after paying his fee.

What Can We Do with this Expression?

We know the expression $m - \$25$ represents our problem if m equals the money earned and $25 is the camping fee. We now have an expression that can be used for any amount of money that Seth earns.

What Can We Do with this Expression? (cont.)

Question: What is the value of the expression if $m = \$79$?

Work: If $m = \$79$, then $m - \$25 = \$79 - \$25$.   Substitute $79 for m.
$\$79 - \$25 = \$54$ Subtract.

Answer: The value is $54.

Understand: If Seth earns $79, then he has $54 left after paying the fee.

Expressions in a Table

Look at each triangle. Each has the one side length of 50 units and a second side length of 40 units. The third side length is different for each triangle. The perimeter is the sum of the three sides (the three sides added together).

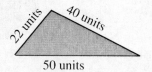 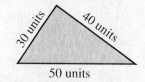 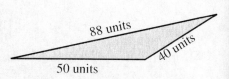

Notice that the third side changes on each triangle. In each case the perimeter equals $40 + 50 +$ the third side. Since the first two sides always add up to 90, the perimeter is always $90 +$ the third side. If we let $s =$ the third side, then $90 + s$ can be used to find the perimeter.

We can create a table showing the perimeter for various side lengths. In each case, we substitute the value for s into the expression $90 + s$. We then work the addition to find the perimeter.

Side length (s)	Showing work ($90 + s$)	Perimeter ($90 + s$)
22	90 + 22	112
30	90 + 30	120
88	90 + 88	178

Side length (s)	22	30	88
Perimeter ($90 + s$)	112	120	178

Expressions in Our Daily Lives

Expressions are used in just about any job you can imagine. They show how numbers, variables, and operations work together. How much paint is needed? How much money will be made? How much rocket fuel should be used? Expressions help us understand all these situations.

You Try It

Karen is decorating cookies for a party. She will set 12 cookies aside for a neighbor. Write an expression to show how many cookies she has left for the party. Let $c =$ the number of cookies she makes.

Express It Mathematically

Think of any number. Add 2 to your number, then add 5, then subtract 1, and finally, subtract 6. Is your result equal to the number you started with? It should be. If we write n in place of your number, your calculations can be written as:

$$n + 2 + 5 - 1 - 6.$$

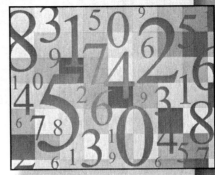

Basic Facts

A **mathematical expression** is any combination of **numbers**, **variables**, and **operations**. Consider these examples:

$$2 + 4 \qquad 2{,}184 - 3 \qquad 7 + 16 \div 4 - 2 \qquad 64 \div 2 - 3$$

Remember that a variable is any symbol used to represent a number. Here are some examples of variables:

$$b, \text{ where } b = 2 \qquad \square, \text{ where } \square = 7 \qquad r, \text{ where } r = 4$$

Once again, let us consider $n + 2 + 5 - 1 - 6$. This is a mathematical expression that includes a variable, numbers, and operations. The variable is n, the operations are $+$ and $-$, and the numbers are 2, 5, 1, and 6.

Writing Expressions

Seth is earning money that he plans to spend on a camping trip. He must pay a $25 fee for the location where he plans to camp regardless of how much money he earns. Write an expression showing the amount of money he has remaining after paying his fee.

Consider the following expressions. Each of these expressions shows Seth's possible earned money minus the fee.

$m - \$25$	
$\$27 - \$25$	What stays the same?
$\$35 - \$25$	*The subtraction of $25 stays the same.*
$\$42 - \$25$	What changes?
$\$67 - \$25$	*The first number changes.*

The variable m represents the number that changes. The variable m can be written in place of the $27, the $35, the $42, and the $67. The expression $m - \$25$ represents the money Seth will have left after paying his fee regardless of how much he earns.

What Can We Do with this Expression?

We know the expression $m - \$25$ represents our problem if m equals the money earned and $25 is the camping fee. We now have an expression that can be used for any amount of money that Seth happens to earn.

What Can We Do with this Expression? (cont.)

Question: What is the value of the expression if $m = \$79$?
Work: If $m = \$79$, then $m - \$25 = \$79 - \$25$. Substitute $79 for m.
$\$79 - \$25 = \$54$ Subtract.
Answer: The value is $54.
Understand: If Seth earns $79, then he will have $54 left over after paying the fee for his site.

Expressions in a Table

Consider the following set of triangles. Each has the one side length of 50 units and a second side length of 40 units. The third side length is different for each triangle. The perimeter is the sum of the three sides (the three sides added together).

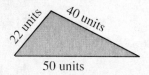 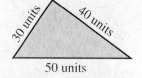 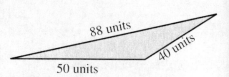

Notice that the third side changes on each triangle. In each case, the perimeter equals $40 + 50 +$ the third side. Since the first two sides always add up to 90, the perimeter is always $90 +$ the third side. If we let $s =$ the third side, then $90 + s$ can be used to find the perimeter.

We can create a table showing the perimeter for various side lengths. In each case, we substitute the value for s into the expression $90 + s$, and then work the addition to find the perimeter.

Side length (s)	Showing work ($90 + s$)	Perimeter ($90 + s$)
22	90 + 22	112
30	90 + 30	120
88	90 + 88	178

Side length (s)	22	30	88
Perimeter ($90 + s$)	112	120	178

Expressions in Our Daily Lives

Expressions show how numbers, variables, and operations work together such as how much paint is needed for a project, how much money will be made in a business venture, or how much rocket fuel should be used to launch a particular spacecraft. Expressions help us understand all these situations.

You Try It

Karen is decorating cookies for a party, and plans to set 12 cookies aside for a neighbor. Write an expression to show how many cookies she has left for the party after giving some to her neighbor. Let $c =$ the total number of cookies she makes.

Expressing More...Mathematically

Who uses mathematical expressions? Many people! People who work in stores do. Even people who work in costume shops do!

A costume shop gets sent 12 boxes. Each box has 6 masks. How many masks did the store get?

Basic Facts

What is a **mathematical expression**? Do you remember? It is any group of **numbers**, **variables**, and **operations**. You have worked with them before. Some use addition. Some use subtraction. You have worked with those. They can use multiplication, too. They can even use division.

How did you solve the problem? You may have written 6×12. This would show that the store got 72 masks. You just wrote an expression.

Let's say the shop gets 12 boxes one week. The next week, they get 5 boxes. Then they get 17 boxes the next week. The last week, they get 10 boxes. We can show the number of masks they got:

6×12 Week 1, with 12 boxes

6×5 Week 2, with 5 boxes

6×17 Week 3, with 17 boxes

6×10 Week 4, with 10 boxes

Each time we show $6 \times$ *the number of boxes*.

A variable is a symbol. It is used in place of a number.

Let $n =$ the number of boxes the store got. Then, $6 \times$ *the number of boxes* may be written as $6 \cdot n$. Or it can be written $6(n)$, or $6n$. Each is an expression. So are the others. Each shows the number of masks when n boxes are present.

Using Tables

Tables can show expressions. You have seen them used for addition. They can be used for subtraction. And for multiplication. And for division, too. It does not matter what the operation is. We use **substitution** to solve. We are given a value for *n*. We use that value *in place of n*.

Number of Boxes (*n*)	1	2	3	10
Showing Work 6•*n*	6(1)	6(2)	6(3)	6(10)
Number of Masks (6*n*)	6	12	18	60

or

Number of Boxes (*n*)	1	2	3	4	5	10	12	15	17	101
Number of Masks (6*n*)	6	12	18	24	30	60	72	90	102	606

Determining Expressions

Look at the table below. Let the input be *k*. What goes in the output?

Input	3	6	15	21	30	60	69	300	*k*
Output	1	2	5	7	10	20	13	100	

You need to figure something out. You need to know what to do to the input to get the output.

When	Input	3	We know 3 − 2 = 1
Then	Output	1	We know 3 ÷ 3 = 1

When	Input	6	We know 6 − 4 = 2
Then	Output	2	We know 6 ÷ 3 = 2

What did you notice? Both times *input ÷ 3 = output*. Check the other pairs. Try 15 and 5. See that 15 ÷ 3 = 5. Check 60 and 20. See that 60 ÷ 3 = 20. Check each input/output pair. We see that *input ÷ 3 = output* every time. Let *k* be the input. Then *k* ÷ 3 goes in the output. It goes in the table. Write *k* ÷ 3 in the blank.

Expressions in Our Daily Lives

How much medicine do you need? What if your mom is sick? How much does she need? Doctors use expressions. This helps them know. This keeps people safe. They find a person's weight. This tells them how much the person needs. Sometimes a weight is given in kilograms. It must be changed to pounds. Expressions help with these conversions.

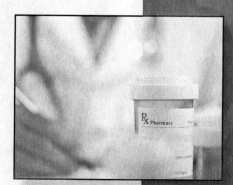

You Try It

Fill in the table.

Input	2	3	5	8	9	10	12	*m*
Output	12	18	30	48	54			

Expressing More...Mathematically

Who uses mathematical expressions? Many people! Even people who work in costume shops do!

A costume shop gets sent 12 boxes. Each box has 6 masks. How many masks did the store get?

Basic Facts

What is a **mathematical expression**? It is any group of **numbers**, **variables**, and **operations**. You have worked with expressions before. Some use addition. And you have used some that had subtraction. They can use multiplication, too. They can even use division.

How did you solve the problem above? You may have written 12 × 6. This would let you know that the store got 72 masks. You just wrote an expression.

Let's say the shop gets 12 boxes one week. The next week, they get 5 boxes. The third week, they get 17. Finally, the fourth week they get 10 boxes.

We can show the number of masks they got:

6 × 12   Week 1, with 12 boxes

6 × 5   Week 2, with 5 boxes

6 × 17   Week 3, with 17 boxes

6 × 10   Week 4, with 10 boxes

Each time we show 6 × *the number of boxes*.

A variable is a symbol. It is any symbol used to show a number.

Let n = the number of boxes the store got. Then 6 × *the number of boxes* is 6n. Or it can be written 6 • n, 6(n), or 6n. Each is an expression. Each shows the number of masks when n boxes are present.

Using Tables

Tables can show expressions. You have seen them used for addition. They can be used for subtraction. They can be used for multiplication, too. And they can be used for division. It does not matter what the operation is. We use **substitution** to solve. We are given a value for *n*. We use that value *in place of n*.

Number of Boxes (*n*)	1	2	3	10
Showing Work 6•*n*	6(1)	6(2)	6(3)	6(10)
Number of Masks (6*n*)	6	12	18	60

or

Number of Boxes (*n*)	1	2	3	4	5	10	12	15	17	101
Number of Masks (6*n*)	6	12	18	24	30	60	72	90	102	606

Determining Expressions

Look at the table below. Let the input be *k*. What expression goes in the output?

Input	3	6	15	21	30	60	69	300	*k*
Output	1	2	5	7	10	20	13	100	

Figure out what you need to do to the input to get the output.

When	Input	3	We know 3 − 2 = 1
Then	Output	1	We know 3 ÷ 3 = 1

When	Input	6	We know 6 − 4 = 2
Then	Output	2	We know 6 ÷ 3 = 2

What did you notice? Both times *input* ÷ 3 = *output*. Check the other pairs. In checking 15 and 5. We see that 15 ÷ 3 = 5. In checking 60 and 20, we see that 60 ÷ 3 = 20. If we check each input/output pair, we see that *input* ÷ 3 = *output* in every case. So if *k* is the input, then *k* ÷ 3 goes in the table for the output. Write *k* ÷ 3 in the blank.

Expressions in Our Daily Lives

How much medicine is needed? Doctors use expressions to help them know. Sometimes they must multiply. They may take a person's weight times a given number. They do this to find out how much medicine to give. Sometimes a weight is given in kilograms. It must be changed to pounds. Sometimes a height is given in inches. It must be changed to centimeters. Expressions help us know.

You Try It

Complete the table.

Input	2	3	5	8	9	10	12	*m*
Output	12	18	30	48	54			

Expressing More...Mathematically

Did you know people who work in costume shops use mathematical expressions?

A costume shop receives a shipment of 12 boxes. Each box contains 6 masks. How many masks did the store receive?

Basic Facts

Remember that a **mathematical expression** is any group of **numbers**, **variables**, and **operations**. You know expressions use addition and subtraction. Expressions can also use multiplication. They can even use division.

Did you write 12 × 6 to see that 72 masks were received? You wrote an expression.

Let's say the shop gets 12 boxes the first week. The second week, the shop gets 5 boxes. The third week, the shop gets 17 boxes. The fourth week, the shop gets 10 boxes.

We can show the number of masks received with:

6×12 Week 1, with 12 boxes

6×5 Week 2, with 5 boxes

6×17 Week 3, with 17 boxes

6×10 Week 4, with 10 boxes

Each time we show $6 \times$ *the number of boxes.*

Remember that a variable is any symbol used to show a number.

If we let n = the number of boxes received, $6 \times$ *the number of boxes* may be rewritten as *6n,* or $6 \cdot n$, or $6(n)$. The expression 6n shows the number of masks received when n boxes are received.

Using Tables

Addition and subtraction expressions can be used in tables. Multiplication and division expressions can also be used in tables. No matter what the operation may be, we use **substitution** to solve. We are given a value for *n*, and we use that value *in place of n*.

Number of Boxes (*n*)	1	2	3	10
Showing Work 6•*n*	6(1)	6(2)	6(3)	6(10)
Number of Masks (6*n*)	6	12	18	60

or

Number of Boxes (*n*)	1	2	3	4	5	10	12	15	17	101
Number of Masks (6*n*)	6	12	18	24	30	60	72	90	102	606

Determining Expressions

Look at the table below. What expression goes in the output when the input is *k*?

Input	3	6	15	21	30	60	69	300	*k*
Output	1	2	5	7	10	20	13	100	

What do you need to do to the input to get the output.

When	Input	3	We know 3 − 2 = 1
Then	Output	1	We know 3 ÷ 3 = 1

When	Input	6	We know 6 − 4 = 2
Then	Output	2	We know 6 ÷ 3 = 2

Did you notice that both times *input ÷ 3 = output*? Check the other pairs. When we check 15 and 5, we see that 15 ÷ 3 = 5. When we check 60 and 20, we see that 60 ÷ 3 = 20. In fact, if we check each input/output pair, we see that *input ÷ 3 = output* in every case. So if *k* is the input, then *k ÷ 3* goes in the table for the output. Write *k ÷ 3* in the blank.

Expressions in Our Daily Lives

How much medicine do we take? Doctors use expressions to answer that question. Sometimes they need to multiply a person's weight with a dose amount. Then they know how much medicine to give. Sometimes we change a weight in kilograms to pounds. Sometimes we change a height in inches to centimeters. Expressions help with these tasks.

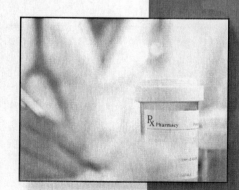

You Try It

Complete the table.

Input	2	3	5	8	9	10	12	*m*
Output	12	18	30	48	54			

Expressing More...Mathematically

Did you know people who work in costume shops use mathematical expressions?

A costume shop receives a shipment of 12 boxes. Each box contains 6 masks. How many masks did they receive in all?

Basic Facts

Remember that a **mathematical expression** is any group of **numbers**, **variables**, and **operations**. You have worked with expressions that use addition and subtraction, but expressions can also use multiplication and division.

You may have written 12×6 to figure out that 72 masks were received. You just wrote an expression.

Let's say the shop gets 12 boxes the first week. The second week the shop gets 5 boxes. The third week the shop gets 17 boxes, and finally, the fourth week, the shop gets 10 boxes.

We can show the number of masks received with these expressions:

6×12 Week 1, with 12 boxes

6×5 Week 2, with 5 boxes

6×17 Week 3, with 17 boxes

6×10 Week 4, with 10 boxes

Each time we show $6 \times$ *the number of boxes.*

Remember that a variable is any symbol used to represent a number.

If we let n = the number of boxes received, $6 \times$ *the number of boxes* may be rewritten as *6n,* or $6 \cdot n$, or $6(n)$. The expression $6n$ shows the number of masks received when n boxes are received.

Using Tables

Addition and subtraction expressions can be used in tables. Multiplication and division expressions can also be used in tables. No matter what the operation may be, we use **substitution** to solve the expression. We are given a value for n and we use that value *in place of n*.

Number of Boxes (n)	1	2	3	10
Showing Work $6 \cdot n$	6(1)	6(2)	6(3)	6(10)
Number of Masks ($6n$)	6	12	18	60

or

Number of Boxes (n)	1	2	3	4	5	10	12	15	17	101
Number of Masks ($6n$)	6	12	18	24	30	60	72	90	102	606

Determining Expressions

Consider the table below. What expression goes in the output when the input is k?

Input	3	6	15	21	30	60	69	300	k
Output	1	2	5	7	10	20	13	100	

Determine what you need to do to the input to get the output.

When	Input	3	We know 3 − 2 = 1
Then	Output	1	We know 3 ÷ 3 = 1

When	Input	6	We know 6 − 4 = 2
Then	Output	2	We know 6 ÷ 3 = 2

Did you notice that both times *input* ÷ *3* = *output*? Check the other pairs based upon the data available.. In checking 15 and 5, we see that 15 ÷ 3 = 5, and in checking 60 and 20, we see that 60 ÷ 20. In fact, if we verify each input/output pair, we see that *input* ÷ *3* = *output* in every case. So if k is the input, then k ÷ 3 goes in the table for the output. Write k ÷ 3 in the blank.

Expressions in Our Daily Lives

How much medicine is required to treat a sick patient? Doctors use expressions to determine those amounts. Sometimes they must multiply a person's weight with an established dose in order to determine how much medicine to give. Sometimes a weight in kilograms must be converted to pounds, or a height in inches must be converted to centimeters. Expressions help with each of those tasks.

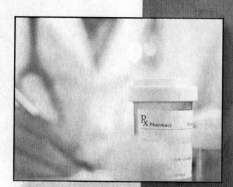

You Try It

Complete the table.

Input	2	3	5	8	9	10	12	m
Output	12	18	30	48	54			

#50716—Leveled Texts for Mathematics: Algebra and Algebraic Thinking © Shell Education

Many Ways to Look at It

Look at the map. What is at (C, 1)? What is at (E, 4) and (H, 2)?

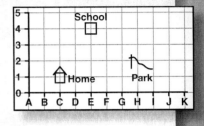

We find a house at (C, 1). There is a school at (E, 4). There is a park at (H, 2). We find these on the **grid**. And, we find points on a **coordinate plane**.

Basic Facts

What you see here is a coordinate plane.

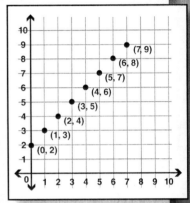

It is made with 2 number lines. One makes the **horizontal axis**. The other makes the **vertical axis**.

Each point is labeled. The label tells us where the point is. We use an **ordered pair** to show this. This follows the order (horizontal coordinate, vertical coordinate). It is like the input and output in a function table. Ordered pairs are written as (input, output).

Look at the pair (5, 7). We were given the coordinates. But, what if we were not? We could find them. We start at the point. Then we look down. We find the matching input of 5. We then start again at the point. And we look to the left to find the output of 7.

Check each point above. Make sure you can name each ordered pair.

Figures, Sequences, and Tables

What comes next in this series of numbers?

2, 3, 4, 5, 6, 7, _____ , ...

If you said 8 is next, you are right. To move from one term to the next, we use the rule "add 1 to the previous number."

What belongs in the remaining blanks of the function table below?

Input	1	2	5	6	7	n
Output	3	4	7	8		

To decide, ask yourself, "What do I do to the input to arrive at the output?" Here, you add 2 to the input to get the output. To fill in the first blank use $7 + 2 = 9$. Write the 9 in your table.

Finally, notice that the rule for finding the output is *input + 2*. The variable n is the input, so the expression $n + 2$ is the output. Write the expression $n + 2$ in the table.

Putting It Together

Look at the coordinate plane, the sequence, and the function table from the other examples. Every one comes from the same information. But the information is shown in different ways. When you show the same information in different ways, it is called using **multiple representations**.

- Look at the first ordered pair on the coordinate plane. It is (0, 2).
- If you call the first number in the sequence "term zero," then the 0th term is 2.
- If you use the expression $n + 2$ from the table, then when the input is 0, the output is 2.

Multiple Representations in Our Daily Lives

How are discoveries made? They start with observations. Patterns might be seen. What helps someone see the patterns? It might help to write information into a table. Maybe we need to see the information on a coordinate plane. The pattern might lead to a formula. A formula can show how different parts are related. Formulas use mathematical expressions.

Patterns, graphs, and expressions help us understand the world. They help us understand a falling apple. We can understand the movement of planets. We can understand the universe around us.

You Try It

Show the information below in three ways.

- Use a graph.
- Use a table.
- Write an expression.

Look at the ordered pairs. Fill in the blanks.

Use the coordinate plane to the right. Fill in the missing information. Look at what you wrote. What goes in the last two spaces? Fill those in. Make drawings to show the first five terms of the sequence 0, 2, 4, 6,…

Input	0	1	2		4	11	n
Output	0			6			

Many Ways to Look at It

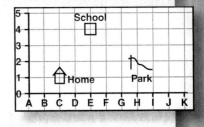

Look at the map. What is found at (C, 1)? What is found at (E, 4) and (H, 2)?

We can find a house at (C, 1). There is a school at (E, 4) and a park at (H, 2). We can find these things on the **grid**. And, we can find points on a **coordinate plane**.

Basic Facts

What you see here is a coordinate plane.

It is made with a number line. It makes the **horizontal axis**. It is made with another number line. It makes the **vertical axis**.

The position of each point is labeled. We use an **ordered pair**. This follows the order (horizontal coordinate, vertical coordinate). It is like the input and output in a function table. Ordered pairs are written as (input, output).

Look at the ordered pair (5, 7). We were given the specific coordinates. But, what if we were not? We could figure them out. We start at the point. Then we look down to the matching input of 5. We then start again at the point. And we look over to the left to find the matching output of 7.

Check each point on the coordinate plane above. Make sure you can name each ordered pair.

Figures, Sequences, and Tables

What comes next in this sequence of numbers?

$$2, 3, 4, 5, 6, 7, \underline{}, \ldots$$

If you said 8 is the next number, you are right. The rule to move from one term to the next is "add 1 to the previous number."

What belongs in the remaining blanks of the function table below?

Input	1	2	5	6	7	n
Output	3	4	7	8		

Remember that when deciding how to fill in a function table, you must ask yourself, "What do I do to the input to arrive at the output?" In this case, notice that you add 2 to the input to arrive at the output. So, to fill in the first blank use $7 + 2 = 9$. Write the 9 in your table.

Finally, notice that the rule for finding the output is *input* + 2. The variable *n* is the input, so the expression $n + 2$ is the output. Write the expression $n + 2$ in the table.

Putting It Together

Look at the coordinate plane from the previous example. Now look at the sequence. And, look at the function table. Every one of these comes from the same information. But, they show that information in different ways. This is called using **multiple representations**.

- Look at the first ordered pair on the coordinate plane. It is (0, 2).
- If you call the first number in the sequence "term zero," then the 0th term is 2.
- If you use the expression $n + 2$ from the table, then when the input is 0, the output is 2.

Multiple Representations in Our Daily Lives

How are discoveries made? They start with observing events. Patterns might be seen. But what helps someone see the patterns? It might help to write the information into a table. Maybe the information needs to be shown on a coordinate plane. The pattern might lead to a formula. A formula can show how different parts of what is seen relate. Formulas use mathematical expressions.

Patterns, graphs, expressions—they all can help us understand our world. They help us understand a falling apple. They help us see the movement of the planets. They can help us notice things. They help us understand the universe.

You Try It

Represent the information below in three ways:

- Use a graph.
- Use a table.
- Write an expression.

Look at the ordered pairs. Fill in the blanks.

Use the coordinate plane to the right. Fill in the missing information. Then, use what you have filled in. Figure out the last two spaces. Then fill them in. Write an expression. Show the first five terms of the sequence 0, 2, 4, 6,…

Input	0	1	2		4		n
Output	0			6		11	

Many Ways to Look at It

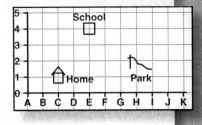

Look at the map. What is found at (C, 1)? What is found at (E, 4) and (H, 2)?

Just as we can locate a house at (C, 1), a school at (E, 4) and a park at (H, 2) on a **grid**, we can locate points on a **coordinate plane**.

Basic Facts

What you see here is a coordinate plane.

It is made with a number line that forms the **horizontal axis**. And it is made with a number line that forms the **vertical axis**.

The location of each point is labeled with an **ordered pair**. This follows the order (horizontal coordinate, vertical coordinate). It is similar to the input and output in a function table. Ordered pairs are written as (input, output).

Look at the ordered pair (5, 7). Even if we were not given the specific coordinates, we could figure them out. We start at the point and look down to the matching input of 5. We then start again at the point and look over to the left to find the matching output of 7.

Check each point on the coordinate plane above and make sure you can name each ordered pair.

Figures, Sequences, and Tables

What comes next in this sequence of numbers?

$$2, 3, 4, 5, 6, 7, \underline{}, \ldots$$

If you said 8 is the next number, you are right. The rule to move from one term to the next is "add 1 to the previous number."

What belongs in the remaining blanks of the function table below?

Input	1	2	5	6	7	n
Output	3	4	7	8		

Remember that when deciding how to fill in a function table, you must ask yourself, "What do I do to the input to arrive at the output?" In this case, notice that you add 2 to the input to arrive at the output. So, to fill in the first blank use $7 + 2 = 9$. Write the 9 in your table.

Finally, notice that the rule for finding the output is *input + 2*. The variable n is the input, so the expression $n + 2$ is the output. Write the expression $n + 2$ in the table.

81

Putting It Together

Look at the coordinate plane, the sequence, and the function table from the previous examples. Every one of these comes from the same information, but they show that information in different ways. When you show the same information in different ways, it is called using **multiple representations**.

- Look at the first ordered pair on the coordinate plane. It is (0, 2).
- If you call the first number in the sequence "term zero," then the 0th term is 2.
- If you use the expression $n + 2$ from the table, then when the input is 0, the output is 2.

Multiple Representations in Our Daily Lives

How are discoveries made? They start with observing events. Patterns might be seen, but what helps someone see the patterns? It might help to write the information into a table. Maybe the information needs to be shown on a coordinate plane. The pattern might lead to a formula. A formula can show how different parts of what is seen are related. Formulas use mathematical expressions.

Patterns, graphs, expressions—they all can help us understand our world. They help us understand a falling apple or the movement of the planets. They can help us understand the universe around us.

You Try It

Represent the information below in three different ways:

- Use a graph.
- Use a table.
- Write an expression.

Fill in the blanks for the ordered pairs below.

Use the coordinate plane below to fill in the missing information. Then, use what you have filled in to figure out and fill in the last two spaces. Write an expression that shows the first five terms of the sequence 0, 2, 4, 6,…

Input	0	1	2		4	11	n
Output	0			6			

Many Ways to Look at It

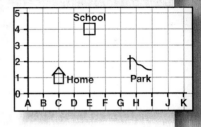

Look at the map. What is located at (C, 1)? What is at coordinates (E, 4) and (H, 2)?

Just as we can locate a house at (C, 1), a school at (E, 4) and a park at (H, 2) on a **grid**, we can locate points on a **coordinate plane**.

Basic Facts

What you observe here is a coordinate plane.

It is constructed with number lines that form the **horizontal axis** and the **vertical axis**.

The location of each point is labeled with an **ordered pair**. This follows the order (horizontal coordinate, vertical coordinate). It is similar to the input and output in a function table. Ordered pairs are written as (input, output).

Look at the ordered pair (5, 7). Even if we were not given the specific coordinates, we could determine what they are. We start at the point and look down to the matching input of 5. We then start again at the point and look over to the matching output of 7.

Verify each point on the coordinate plane above and make sure you can name each ordered pair.

Figures, Sequences, and Tables

What comes next in this sequence of numbers?

$$2, 3, 4, 5, 6, 7, \_\_\_,...$$

If you said 8 is the next number, you are right. The rule that stipulates how to move from one term to the next is "add 1 to the previous number."

What belongs in the remaining blanks of the function table below?

Input	1	2	5	6	7	n
Output	3	4	7	8		

It is important to note that when deciding how to fill in a function table, you must ask yourself, "What do I do to the input to arrive at the output?" In this case, notice that you add 2 to the input to generate the output. So, to fill in the first blank apply the rule: $7 + 2 = 9$. Write the 9 in your table.

Finally, notice that the rule for finding the output is *input + 2*. The variable n is the input, so the expression $n + 2$ is the output. Write the expression $n + 2$ in the table.

83

Putting It Together

Revisit the coordinate plane, the sequence, and the function table from the previous examples. They display the same information, but in different ways. Expressing the same information in different ways is called using **multiple representations**.

- Look at the first ordered pair on the coordinate plane. It is (0, 2).
- If you call the first number in the sequence "term zero," then the 0th term is 2.
- If you use the expression $n + 2$ from the table, then when the input is 0, the output is 2.

Multiple Representations in Our Daily Lives

Discoveries are fostered by someone's ability to initiate a keen observation. Patterns emerge, but what facilitates someone's ability to see them? Organizing and representing the information into one of the forms discussed organizes information. Translate the information into a table. Represent it on a coordinate plane. Decipher a pattern and apply a formula. A formula can underscore relationships among divergent aspects of the observation. Formulas represent a utilization of mathematical expressions.

Patterns, graphs, expressions—they all can help us understand our world. They give us insight into the forces behind a falling apple or the mysterious movement of the planets. They can enlighten us as to the complexities of the universe around us.

You Try It

Represent this information in three different ways:

- Use a graph.
- Use a table.
- Write an expression.

Fill in the blanks for the ordered pairs below.

Use the coordinate plane below to fill in the missing information. Then, use your findings to determine the missing answers for the last two spaces. Write an expression that shows the first five terms of the sequence 0, 2, 4, 6,…

Input	0	1	2		4	11	n
Output	0			6			

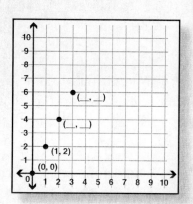

Adding Some Balance

Tonya has one dog. Jamaal has two dogs. Together they have three dogs.

We can write 4 statements. They are all about the numbers of dogs. They make a fact family.

$$1 + 2 = 3 \qquad 3 - 1 = 2$$
$$2 + 1 = 3 \qquad 3 - 2 = 1$$

Basic Facts

An **equation** is a mathematical statement. Each side has a value. Look at the equation $1 + 2 = 3$. The value of "$1 + 2$" is 3. And the value of "3" is 3. Both sides equal 3. They have the same value. This means the equation is true. Look at the fact family above. Each statement is an equation. Both sides are equal.

Think of the two sides as a **balance**. The two sides must have the same value. This lets it stay in balance.

Unknowns in Equations

Sometimes there are things that are not known. Think about this. Martino has 2 birds. Keisha has some birds. Keisha's birds are in a covered cage. We do not know how many she has. Together Martino and Keisha have 7 birds.

What do we know? Martino's birds + Keisha's birds = total number of birds

We are given amounts. This lets us draw what we know. The cage can stand for Keisha's birds.

 + =

Martino's birds   Keisha's birds   Total number of birds

Drawings helps us understand, but it would not make sense to always have to draw what we know. What if Martino owned a pet store? What if he had 400 birds? Instead, we can assign a variable. This will stand for the number of birds Keisha has. We use numbers for things that we know.

Assign a variable.	Let b = number of Keisha's birds
Write an equation.	$2 + b = 7$
Use fact families or you can guess and check.	$2 + b = 7$ is the same as $7 - 2 = b$
Keisha has 5 birds.	$5 = b$

Writing Equations

Evan owns a pet store. He added 7 *new* kinds of fish to the tanks. He now sells 11 kinds.

You may have seen problems like this before. You might be asked, "How many kinds of fish did Evan start with?" You can find the answer. You can subtract. Use $11 - 7$. Evan started with 4 kinds of fish.

This time let's write an equation. First use pictures. It looks like this:

Now let us use variables. You can write it like this.

Assign a variable. Let f = number of kinds of fish

Write an equation. $f + 7 = 11$

Think of fact families. We know that $f + 7 = 11$ has the same value as $11 - 7 = f$. And $11 - 7 = 4$. We now know $f = 4$. So, we know that Evan started with 4 fish.

Addition Equations in Our Daily Lives

You have written equations. You have found the values of variables. And you have used Guess and Check. These can all help you. But what if the numbers are bigger? Or, what if the problems have fractions? What if they use decimals? Equations are important. Many people need them for their jobs. How many things should be sold? What should they cost? Equations help show situations. Need to show a problem? Try an equation. Need to make a decision about an amount? Equations can help.

You Try It

Dakota has to do a science project. She worked 6 hours. She worked some more to finish the project. She worked a total of 13 hours.

- Show this in pictures.
- Assign a variable for the number of more hours worked.
- Write an equation for the problem.
- How many more hours did Dakota work?

Adding Some Balance

Tonya has one dog. Jamaal has two dogs. Together they have three dogs.

We can write 4 statements. They are all about the numbers of dogs. They make a fact family.

$$1 + 2 = 3 \qquad 3 - 1 = 2$$
$$2 + 1 = 3 \qquad 3 - 2 = 1$$

Basic Facts

An **equation** is a mathematical statement. Each side of the equation has a value. In the equation $1 + 2 = 3$, the value of "$1 + 2$" is 3. And the value of "3" is 3. For an equation to be true the values of both sides of the equation must be equal. Look at the fact family above. Each statement is an equation.

The two sides should be thought of as a **balance**. The two sides must be the same in value for the balance to stay equal.

Unknowns in Equations

Sometimes there are things that are unknown in an equation. Think about this situation. Martino has 2 birds. And Keisha has some birds. Keisha's birds are in a covered cage. We are not sure how many birds Keisha has. Together Martino and Keisha have 7 birds.

We know: Martino's birds + Keisha's birds = total number of birds

We are given amounts. This lets us draw what we know. The cage can stand for Keisha's birds.

 + =

Martino's birds   Keisha's birds   Total number of birds

It would not be reasonable to always have to draw what we know. What if Martino owned a pet store? What if he had 400 birds? Instead, we can assign a variable for the number of birds Keisha has. We use numbers for the other information that we know.

Assign a variable.	Let b = number of Keisha's birds
Write an equation.	$2 + b = 7$
Use fact families or you can guess and check.	$2 + b = 7$ is the same as $7 - 2 = b$
Keisha has 5 birds.	$5 = b$

Writing Equations

Evan owns a pet store. He added 7 *new* kinds of fish to his stock. He now sells 11 kinds.

You may have seen problems like this before. You might be asked, "How many kinds of fish did Evan start with?" You find the answer by subtracting $11 - 7$. So, Evan started with 4 kinds of fish.

This time let's write an equation. First use pictures. It looks like this:

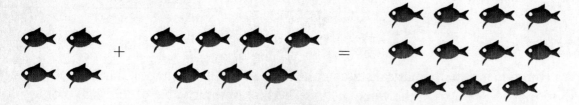

Now let us use variables. Your equation can be written like this.

Assign a variable.　　　　Let f = number of kinds of fish

Write an equation.　　　　$f + 7 = 11$

With fact families we know that $f + 7 = 11$ has the same value as $11 - 7 = f$. We now know $f = 4$. So we know that Evan started with 4 fish.

Addition Equations in Our Daily Lives

You have written equations. You have found the values of variables. And you have used Guess and Check to help you. But what if the numbers are larger? Or, what if the problems have fractions or decimals? Equations are important. They are necessary in many peoples' jobs. How many kinds of items should be sold? What should their price be? Equations help show situations. Need to show a problem? Try an equation. Need to make a decision about an amount? Equations can help.

You Try It

Dakota has worked 6 hours on a science project. Then she worked some more to finish the project. She worked a total of 13 hours.

- Show your situation in pictures.
- Assign a variable for the number of additional hours worked.
- Write an equation for the situation.
- How many additional hours did Dakota work?

Adding Some Balance

Tonya has one dog. Jamaal has two dogs. Together they have three dogs.

We can write 4 statements. They all describe the numbers of dogs; together, they make a fact family.

$$1 + 2 = 3 \qquad 3 - 1 = 2$$
$$2 + 1 = 3 \qquad 3 - 2 = 1$$

Basic Facts

An **equation** is a mathematical statement. Each side of the equation has a value. In the equation $1 + 2 = 3$, the value of "$1 + 2$" is 3 and the value of "3" is 3. For an equation to be true, the values of both side of the equation must be the same. Every statement in the fact family shown above is a true equation.

Think of the two sides of an equation in terms of a **balance**. The two sides must be the same in value in order for the balance to be equal.

Unknowns in Equations

Sometimes we encounter unknowns in equations. Consider this situation: Martino has 2 birds and Keisha has an undetermined number of birds that she keeps in a covered birdcage. Altogether, Martino and Keisha have 7 birds.

We know: Martino's birds + Keisha's birds = total number of birds

We are given amounts so we can draw what we know. The cage can represent Keisha's birds.

 + =

Martino's birds   Keisha's birds   Total number of birds

It would be unreasonable to always have to draw what we know. What if Martino owned a pet store and had 400 birds? Instead, we can assign a variable for the number of birds Keisha has and use numbers for the other pieces of information that we know.

Assign a variable.	Let b = number of Keisha's birds
Write an equation.	$2 + b = 7$
Use fact families or you can guess and check.	$2 + b = 7$ is the same as $7 - 2 = b$
Keisha has 5 birds.	$5 = b$

Writing Equations

Evan owns a pet store. He recently added 7 *new* kinds of fish to his stock. He now sells 11 different kinds of fish.

You have probably seen situations like this before. You might be asked, "How many kinds of fish did Evan start with?" You find the answer by subtracting $11 - 7$. So, Evan started with 4 kinds of fish.

This time let's write an equation for the situation. In pictures, it looks like this:

Using variables, your equation can be written as:

Assign a variable.   Let f = number of kinds of fish

Write an equation. $f + 7 = 11$

With fact families, we know that $f + 7 = 11$ has the same value as $11 - 7 = f$. And $11 - 7 = 4$. Since $f = 4$, we know that Evan started with 4 fish.

Addition Equations in Our Daily Lives

You have written equations, found the values of variables, and used the strategy Guess and Check to help you. But what if the numbers are larger or what if the problems include fractions or decimals? Equations are important and necessary in many peoples' jobs. How many kinds of items should be sold? What should the price be for the items? Equations help represent situations. Need to represent a problem? Try an equation. Need to make a decision about an amount? Equations can help.

You Try It

Dakota has worked 6 hours on a science project. She worked an additional amount of hours to finish the project. She worked a total of 13 hours.

- Show the situation in pictures.
- Assign a variable for the number of additional hours that Dakota worked.
- Write an equation for the situation.
- How many additional hours did Dakota work?

Adding Some Balance

Tonya has one dog and Jamaal has two dogs. Altogether, they have three dogs.

There are four expressions we can write about the numbers of dogs that constitute the same fact family.

$1 + 2 = 3$	$3 - 1 = 2$
$2 + 1 = 3$	$3 - 2 = 1$

Basic Facts

An **equation** is a mathematical statement in which each side of the equal sign has a value that must be equal to the value of the opposite side. In the equation $1 + 2 = 3$, the value of "$1 + 2$" is 3 and the value of "3" is 3. Every statement in the fact family shown above is a true equation because the values of both sides are equal.

Imagine two sides of an equation in terms of a **balance**. The two sides must necessarily be the same in value in order to maintain the balance.

Unknowns in Equations

Often times, there are variables that are unknown in an equation. Consider this scenario: Martino has 2 birds and Keisha has an unknown quantity of birds in a covered cage. We are not sure of the exact amount of Keisha's birds. Martino and Keisha combined have 7 birds, but we do not have any inkling how many birds Keisha is keeping in her covered cage.

We know that Martino's birds + Keisha's birds = total number of birds

We are given some values, so we can illustrate the information that we do know. In this case, the cage can represent Keisha's birds.

 + =

Martino's birds   Keisha's birds   Total number of birds

It is impractical to always have to draw what we know. What if Martino owned a pet store and had 400 birds? We can facilitate the process by assigning a variable for the number of birds Keisha has and use numbers for the remaining discernible information.

Assign a variable.	Let b = number of Keisha's birds
Write an equation.	$2 + b = 7$
Use fact families or you can guess and check.	$2 + b = 7$ is the same as
	$7 - 2 = b$
Keisha has 5 birds.	$5 = b$

Writing Equations

Evan owns a pet store, and recently he upgraded his inventory with 7 *new* varieties of fish. As a result, there is now a grand total of 11 varieties of fish that he sells.

You have probably encountered situations like this before. You might be asked, "How many varieties of fish did Evan start with?" The answer can be discovered by subtracting with 11 – 7, so you know that Evan started with 4 types of fish.

This time let's write an equation for the situation. In pictures, it looks like this:

Using variables, your equation can be written as:

Assign a variable.   Let f = number of varieties of fish

Write an equation. $f + 7 = 11$

With fact families, we know that $f + 7 = 11$ has the same value as $11 - 7 = f$. And $n - 7 = 4$. Since $f = 4$, we know that Evan started with 4 fish.

Addition Equations in Our Daily Lives

You have written equations, found the values of variables, and used the strategy Guess and Check to help you. But what if the numbers are larger or what if the problems include fractions or decimals? Equations are important and necessary in many peoples' jobs. How many kinds of items should be sold? What should the price be for the items? Equations help represent situations. Need to represent a problem? Try an equation. Need to make a decision about an amount? Equations can help.

You Try It

Dakota worked 6 hours on a science project. Then she worked additional hours to finish it She worked a total of 13 hours.

- Show the situation in pictures.
- Assign a variable for the number of additional hours that Dakota has worked.
- Write an equation for the situation.
- How many additional hours did Dakota work?

Keeping the Balance When Taking Away

Look at this astronaut. He has been to the moon. He lost weight while in space. But now he is back on Earth. In one day, he gains 3 more pounds. Now, he weighs 166 lbs. But how much did he weigh on the day he landed? We can use an equation to find out.

Let w = his weight when he landed back on Earth.
$$w + 3 = 166 \text{ lbs.}$$

He weighed 163 pounds on the day he landed back on Earth.

Basic Facts

What is written in the problem above is an equation. It is a mathematical statement. The **values** on each side must be equal. This makes it true.

Think of the two sides. Now, think of a **balance**. We can put the expression $w + 3$ on the left side. We can place 166 on the right side. The two sides have the same value. So the balance will stay equal. The equation is true.

Equations with Subtraction

Equations are useful. They are good tools. We can use them to add. We can use them to subtract, too.

Think about this. A man is going to the moon. On the day of lift-off he weighs 169 pounds. Then he gets back. He weighs less. He weighs 163 pounds. We can show this with an equation. And we can solve it. This will show how much weight he lost.

We can write an equation in words: *lift-off weight – weight lost = return weight*

Assign a variable.	Let w = weight lost
Write an equation.	$169 - w = 163$

Now let's find what w is.
Use fact families.	$169 - 163 = w$
Or, use guess and check.	$6 = w$

Check your answer.
Write the original equation.	$169 - w = 163$
Substitute 6 for w. Then solve.	$169 - 6 = 163$
So, the sides are balanced!	$163 = 163$

The astronaut lost 6 pounds.

Writing Equations

Think about the man going to the moon. It was normal to lose weight for a moon trip. This happened even before lift-off.

The man weighs 2 pounds less on the day of lift-off than during training. He weighs 169 pounds at lift-off. So, we can find how much he weighed in training.

First, let us use words: *training weight − weight lost = lift-off weight*

Assign a variable.	Let t = weight during training
Write an equation.	$t - 2 = 169$ (lift-off weight)

Now let's find the value of t.

Use fact families.	$169 + 2 = t$
Or, you can guess and check.	$t = 171$

You should check your answer.

Write the original equation.	$t - 2 = 169$
Substitute 171 for t. Then solve.	$171 - 2 = 169$
So, the sides do balance!	$169 = 169$

The astronaut weighed 171 pounds.

Subtraction Equations in Our Daily Lives

We just saw a few equations. They show simple subtraction equations. They are very important. They can help with jobs like space travel. Equations can help in meal planning. They help figure out how much oxygen is needed in space. They help choose when to lift off. And they help know when to land. We need equations. They will help in future space travel.

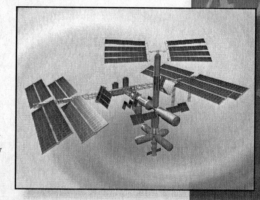

You Try It

Estella has saved money. She wants to visit the Johnson Space Center. It is in Houston, Texas. She spends $42.50 on three things. She buys her ticket, food, and a souvenir. She has $6.25 left. How much money had she initially saved for the trip?

- Write the information as an equation in words.
- Choose a variable for the amount of money that Estella had saved.
- Write an equation for the situation.
- How much money did Estella save for the trip?

Keeping the Balance When Taking Away

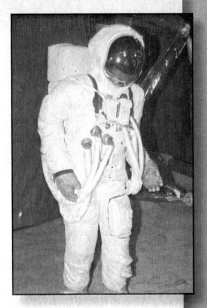

Look at this astronaut. He has been to the moon. He lost weight while in space. But now he is back on Earth. In one day, he gains 3 more pounds. Now, he weighs 166 pounds. But how much did he weigh before he landed? We can use an equation to find out.

Let w = his weight when he landed back on Earth
$$w + 3 = 166$$

He weighed 163 pounds on the day he landed back on Earth.

Basic Facts

What is written in the problem above is an equation. It is a mathematical statement. For it to be true, the **values** on each side must be equal.

Think of the two sides in terms of a **balance**. We can put the expression $w + 3$ on the left side of the balance. We can place 166 on the right side. The two sides have the same value. So the balance will stay equal. The equation is true.

Equations with Subtraction

Equations are useful. They are good tools. We can use them for addition. They are good for subtraction, too.

Think about this. An astronaut weighs 169 pounds. This is on the day of lift-off to the moon. Then he gets back. He weighs less. He weighs 163 pounds. We can show this with an equation. And we can solve it. This lets us find how much weight he lost.

We can write an equation in words: *lift-off weight − weight lost = return weight*

Assign a variable.	Let w = weight lost
Write an equation.	$169 - w = 163$

Now let's find what w is.
Use fact families.	$169 - 163 = w$
Or, you can guess and check.	$6 = w$

You should check your answer.
Write the original equation.	$169 - w = 163$
Substitute 6 for w.	$169 - \mathbf{6} = 163$
So, the sides are balanced!	$163 = 163$

The astronaut lost 6 pounds.

Writing Equations

Recall our astronaut again. It is appropriate to lose weight for a trip to the moon. This happened even before the day of lift-off.

Our astronaut weighs 2 pounds less on the day of lift-off than he did during training. He weighs 169 pounds at lift-off. So, let's determine how much he weighed during training

Use words to make an expression: *training weight – weight lost = lift-off weight*

Assign a variable.	Let t = weight during training
Write an equation.	$t - 2 = 169$

Now let's find the value of t.

Use fact families.	$169 + 2 = t$
Or, you can guess and check.	$t = 171$

You can check your answer. And you should.

Write the original equation.	$t - 2 = 169$
Substitute 171 for t.	$171 - 2 = 169$
So, the sides do balance!	$169 = 169$

The astronaut weighed 171 pounds.

Subtraction Equations in Our Daily Lives

We just saw a few equations. They just start to show subtraction equations. These are important. They show all the ways they help in jobs like space travel. Equations can help in meal planning. They help figure out how much oxygen is needed in space. They help choose when to lift off. And they help know when to land. We need equations. They will help the future in space travel.

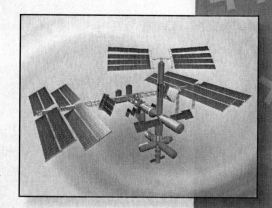

You Try It

Estella has saved money. She wants to visit the Johnson Space Center. It is in Houston, Texas. She spends $42.50 on three things. She buys her ticket, food, and a souvenir. She has $6.25 left. How much money had she initially saved for the trip?

- Write the information as an equation in words.

- Choose a variable for the amount of money that Estella had saved.

- Write an equation for the situation.

- How much money did Estella save for the trip?

Keeping the Balance When Taking Away

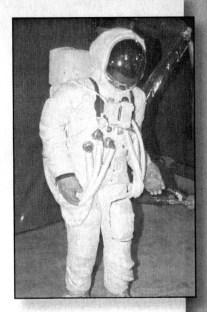

Imagine an astronaut returning to Earth from the moon. He lost weight during the moon mission, but he gains three pounds just one day after return. He weighs 166 pounds after gaining the weight. We can use an equation to find his weight when he landed on Earth.

Let w = weight of astronaut when he landed back on Earth
$$w + 3 = 166$$

Our astronaut weighed 163 pounds when he first landed back on Earth.

Basic Facts

What is written in the problem above is an equation. It is a mathematical statement. For it to be true, the **values** on each side of the equation must be equal.

The two sides of an equation may be thought of in terms of a **balance**. We can place the expression $w + 3$ on the left side of the balance. We can place 166 on the right side. The two sides have the same value, so the balance will stay equal. The equation is true.

Equations with Subtraction

Equations are useful tools for working with addition. They are useful with subtraction, as well.

Think about this situation: An astronaut weighs 169 pounds on the day of lift-off to the moon. On the day of return, he weighs 163 pounds. We can show this with an equation. We can solve our equation to find out the amount of weight he lost.

We can write an equation in words: *lift-off weight − weight lost = return weight*

Assign a variable.	Let w = weight lost
Write an equation.	$169 - w = 163$

Now let's find what w is equal to.
Use fact families.	$169 - 163 = w$
Or, use guess and check.	$6 = w$

You should check your answer.
Write the original equation.	$169 - w = 163$
Substitute 6 for w. Then solve.	$169 - 6 = 163$
So, the sides are balanced!	$163 = 163$

The astronaut lost 6 pounds.

Writing Equations

Let's think about our astronaut again. During the moon missions it is typical for the astronauts to lose some weight, even before the day of lift-off. Our astronaut weighs 2 pounds less on the day of lift-off than he did during training. If he weighs 169 pounds at lift-off, we can find how much he weighed during training.

Our equation in words: *training weight – weight lost = lift-off weight*

Assign a variable.	Let t = weight during training
Write an equation.	$t - 2 = 169$

Now let's find the value of t.

Use fact families.	$169 + 2 = t$
Or, you can guess and check.	$t = 171$

You can and should check your answer.

Write the original equation.	$t - 2 = 169$
Substitute 171 for t. Then solve.	$171 - 2 = 169$
So, the sides do balance!	$169 = 169$

The astronaut weighed 171 pounds.

Subtraction Equations in Our Daily Lives

The few equations you have just seen merely scratch the surface of how subtraction equations contribute to space travel. Equations can determine meal planning for astronauts; they establish the minimum amount of oxygen that should be available at any given time; and they are used when deciding lift-off and landing time frames. Whatever the future of space travel may hold, we will need equations.

You Try It

Estella has saved money to visit the Johnson Space Center in Houston, Texas. She spends $42.50 on her ticket, food, and a souvenir. She has $6.25 when she leaves.

How much money had she initially saved for the trip?

- Write the information as an equation in words.
- Assign a variable for the amount of money that Estella had saved.
- Write an equation for the situation.
- How much money did Estella save for the trip?

Keeping the Balance When Taking Away

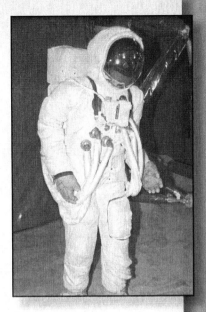

Imagine an astronaut returning to Earth after a mission to the moon. He lost weight during his time in space, but he regains 3 pounds just one day after returning to Earth. We know that after regaining the weight he weighs in at 166 pounds, so we can determine an equation to figure his weight on the day he initially touched back down on Earth.

Let w = weight of astronaut when he first returned to Earth
$$w + 3 = 166$$

Our astronaut weighed 163 pounds when he first landed back on Earth.

Basic Facts

What is demonstrated in the above-mentioned problem is an equation. It is a mathematical statement that is true when the **values** on each side are balanced and equal.

The two sides of an equation may be thought of in terms of a **balance**. We can place the expression $w + 3$ on the left side of the balance, and put 166 on the right side. The two sides have to maintain the same value to keep the balance equal and this is how we know that the equation is true.

Equations with Subtraction

We have seen that equations can be powerful tools for working with addition. They are equally invaluable for situations that involve subtraction.

For example, consider this: An astronaut weighs 169 pounds on the day of lift-off to the moon, but on the day of return, he weighs only 163 pounds. We can show this with an equation, and then we can solve our equation to find out the amount of weight he lost.

We can write an equation in words: *lift-off weight – weight lost = return weight*

Assign a variable.	Let w = weight lost
Write an equation to show the problem.	$169 - w = 163$

Now let's find what w is equal to.

Use fact families.	$169 - 163 = w$
Or, use guess and check.	$6 = w$

You should verify your answer.

Write the original equation.	$169 - w = 163$
Substitute 6 for w. Then solve.	$169 - 6 = 163$
So, the sides are balanced!	$163 = 163$

The astronaut lost 6 pounds.

Writing Equations

Let's reconsider our astronaut. During missions to the moon, it is typical for the astronauts to lose some weight, even before the day of lift-off.

Suppose our astronaut weighs 2 pounds less on the day of lift-off than he did during training. If he weighs 169 pounds at lift-off, we can determine the unknown of how much he weighed during training.

Our equation in words looks like this: *training weight − weight lost = lift-off weight*

| Assign a variable for the missing data. | Let t = weight during training |
| Write an equation that describes the situation. | $t - 2 = 169$ |

Now let's find the value of t.

| Use fact families. | $169 + 2 = t$ |
| Or, you can guess and check. | $t = 171$ |

You can and should verify your answer.

Write the original equation.	$t - 2 = 169$
Substitute 171 for t.	$171 - 2 = 169$
So, the sides do balance!	$169 = 169$

The astronaut weighed 171 pounds during training.

Subtraction Equations in Our Daily Lives

The equations you have just experienced barely show the important role that subtraction equations play in space travel. Equations can help in meal planning for astronauts, they help when calculating the amount of oxygen that needs to be available in space, and they are used in deciding when to lift off and when to land. We obviously will continue to use equations, no matter what the future of space travel may hold.

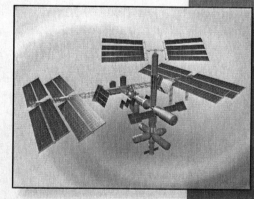

You Try It

Estella has saved money to visit the Johnson Space Center in Houston, Texas. She spends $42.50 on her ticket, food, and a souvenir. After, she has $6.25 remaining. How much money had she initially budgeted for the trip?

- Write the information as an equation in words.

- Assign a variable for the amount of money that Estella had saved.

- Write an equation for the situation.

- How much money did Estella save for the trip?

The Equations Keep Multiplying

How many eyes do you see? Did you notice that you could use multiplication to help answer the question? We could count every eye one by one. Or, we can multiply. When we multiply we find that 5 columns × 4 rows = 20 eyes.

Basic Facts

The statement $5 \times 4 = 20$ is an equation. It is a mathematical statement. The equation is true. We know because the values on each side of the equation are equal.

The statement $5 \times k = 20$ is an equation. It is a mathematical statement. The variable is k. The k stands for the unknown value. We do not use × for multiplication when variables are used. It might confuse readers. Instead, we write $5 \cdot k = 20$. When you figure out that $k = 4$, you have solved the equation. You have found the value for the variable that makes the equation true.

Equations with Multiplication

Think of setting up a rectangular soccer field. You know that the total area of the field is 6,000 square feet. The length is 100 feet. What is the width?

We can use an equation. It will show the situation. We know that the field is a rectangle. We can multiply the length by the width of a rectangle to find the area.

We can write it in words: *length × width = area*.

Assign a variable.	Let w = width
Write an equation.	$100 \cdot w = 6{,}000$

Now let's find the value of w.

Use fact families.	$6{,}000 \div 100 = w$
Or, guess and check.	$60 = w$

You should check your answer.

Write the original equation.	$100 \cdot w = 6{,}000$
Substitute 60 for w. Solve.	$100 \cdot \mathbf{60} = 6{,}000$
So, the sides are balanced!	$6{,}000 = 6{,}000$

The width is 60 feet.

Equations and Formulas

Formulas tell us how values are related.

Here is the formula. It is for the volume of this shape.

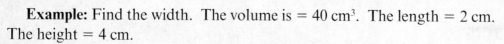

volume = length × width × height

We put what we know into the formula. This lets us solve to find the missing value.

Example: Find the width. The volume is = 40 cm³. The length = 2 cm. The height = 4 cm.

Decide what you must find.	Let w = width
Write the formula.	volume = length × width × height
Put in what you know.	40 cm³ = 2 cm • w • 4 cm
The 8 cm comes from 2 cm × 4 cm.	40 cm³ = 8 cm • w
Use fact families. Or, guess and check.	40 cm³ ÷ 8 cm = w
The width is 5 cm.	5 = w

You can check your answer. You should do this.

Write the formula.	volume = length × width × height
Put in what you know.	40 = 2 • 5 • 4
The sides do balance!	40 = 40
The width is 5 cm.	

Multiplication Equations in Our Daily Lives

What happens if you put money in a savings account at the bank and leave it? Years later you could have more money than you put in! Banks pay **interest**. That means you are paid for the money you have in an account. You may pay interest, too. That is when you pay extra money for money you borrow. Banks figure out interest by using equations. Many things use equations. People who save money use them. People buying a house need them. And people who own a business use them, too.

You Try It

A box has a volume = 24 cm³. It has a length of 3 cm and a width of 2 cm. Draw a picture of the box. Show the measurements.

- Which value is missing? Assign a variable for the value.

- Write an equation for the situation by substituting values into volume = length × width × height.

- What is the value of the missing height?

The Equations Keep Multiplying

How many eyes do you see? Did you notice that you could use multiplication to help answer the question? We could count every eye one by one. Or, we can multiply. When we multiply, we find that 5 columns × 4 rows = 20 eyes.

Basic Facts

The statement *5 × 4 = 20* is an equation. It is a mathematical statement. The equation is true. We know because the values on each side of the equation are equal.

The statement *5 × k = 20* is also an equation. It is a mathematical statement. The variable is k. The k stands for the unknown value. Remember that we do not use × for multiplication when variables are used. It might confuse people. Instead, we write $5 \cdot k = 20$. When you figure out that $k = 4$, you have solved the equation. You have found the value for the variable that makes the equation true.

Equations with Multiplication

Think of setting up a rectangular soccer field. You know that the total area of the field is 6,000 square feet. The length is 100 feet. What is the width?

We can use an equation. It will show the situation. We know that the field is a rectangle. We know that multiplying length by width of a rectangle tells us the area.

We can write it in words: *length × width = area*.

Assign a variable.	Let w = width
Write an equation.	$100 \cdot w = 6{,}000$

Now let's find the value of w.

Use fact families.	$6{,}000 \div 100 = w$
Or, guess and check.	$60 = w$

You should check your answer.

Write the original equation.	$100 \cdot w = 6{,}000$
Substitute 60 for w. Solve.	$100 \cdot \mathbf{60} = 6{,}000$
So, the sides are balanced!	$6{,}000 = 6{,}000$

The width is 60 feet.

Equations and Formulas

Formulas tell us how values are related.

Here is the formula. It is for the volume of this shape.

volume = length × width × height

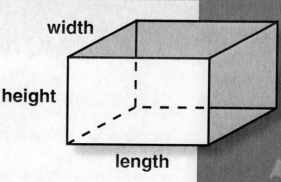

We put what we know into the formula. This lets us solve to find the missing value.

Example: Find the width. The volume is = 40 cm^3. The length = 2 cm. The height = 4 cm.

Decide what you must find.	Let w = width
Write the formula.	volume = length × width × height
Put in what you know.	40 cm^3 = 2 cm • w • 4 cm
The 8 cm comes from 2 cm × 4 cm.	40 cm^3 = 8 cm • w
Use fact families. Or, guess and check.	40 cm^3 ÷ 8 cm = w
The width is 5 cm.	5 = w

You can check your answer. You should do this.

Write the formula.	volume = length × width × height
Put in what you know.	40 = 2 • 5 • 4
The sides do balance!	40 = 40
The width is 5 cm.	

Multiplication Equations in Our Daily Lives

Did you know that if you put money in a savings account at the bank and left it alone, many years later you could have more money than you originally put in? Banks pay **interest**. Earning interest means the bank offers a fee for the money you have in an account. You pay interest when you pay an extra fee for money you borrow. Banks calculate interest through the use of equations. Other situations call for equations, too. These include saving money, buying a house, and owning a business.

You Try It

A box has a volume = 24 cm^3. It has a length of 3 cm. It has a width of 2 cm. Draw a visual representation of the box. Highlight the measurements.

- Which value is missing? Use a variable for the value.

- Write an equation for the situation. Substitute values into volume = length × width × height.

- What is missing is the value of the height. Can you find it?

The Equations Keep Multiplying

Can you tell at a glance how many eyes there are? Did you consider that multiplication could prove useful to answer that question? Rather than counting every eye one by one, it is more efficient to multiply. By multiplying, we find that 5 columns × 4 rows = 20 eyes.

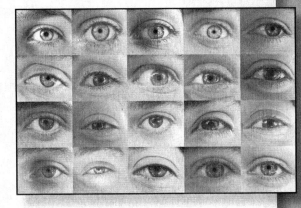

Basic Facts

The mathematical statement 5 × 4 = 20 is an equation. For the equation to be true, the values on each side of the equation must be equal.

The statement $5 \times k = 20$ is also an equation. It is a mathematical statement. The variable is k. The k represents the unknown value. Remember that we do not use × for multiplication when using variables, because that could cause confusion. Instead, we write $5 \cdot k = 20$. When you figure out that $k = 4$, you have solved the equation, which means you know the value of the variable that makes the equation true.

Equations with Multiplication

Imagine you are hired to construct a rectangular soccer field. The total area of the field has to be 6,000 square feet. The length available is 100 feet. Given this information, what would the width have to be?

We can use an equation to assist with this situation. The soccer field is obviously a rectangle. Therefore, we know that we can determine a rectangle's area by multiplying its length by its width.

So in words we know: *length × width = area*.

Assign a variable.	Let w = width
Write an equation.	$100 \cdot w = 6{,}000$

Now let's find the value of w.

Use fact families.	$6{,}000 \div 100 = w$
Or, guess and check.	$60 = w$

You should verify your answer.

Write the original equation.	$100 \cdot w = 6{,}000$
Substitute 60 for w. Solve.	$100 \cdot \mathbf{60} = 6{,}000$
So, the sides are balanced!	$6{,}000 = 6{,}000$

The width is 60 feet.

Equations and Formulas

Formulas tell us how values are related.

The formula for the volume of this figure is:

volume = length × width × height

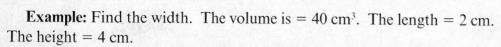

If we start by inserting what we know into the formula, we can solve to find the missing value.

Example: Find the width. The volume is = 40 cm^3. The length = 2 cm. The height = 4 cm.

Decide what you must find.	Let w = width
Write the formula.	volume = length × width × height
Put in what you know.	40 cm^3 = 2 cm • w • 4 cm
The 8 cm comes from 2 cm × 4 cm.	40 cm^3 = 8 cm • w
Use fact families. Or, guess and check.	40 cm^3 ÷ 8 cm = w
The width is 5 cm.	5 = w

You can verify your answer. You should do this.

Write the formula.	volume = length × width × height
Put in what you know.	40 = 2 • 5 • 4
The sides do balance!	40 = 40
The width is 5 cm.	

Multiplication Equations in Our Daily Lives

Did you know that if you put money in a savings account at the bank and left it alone, many years later you could have more money than you first put in? Banks pay **interest**. Earning interest means you are paid for the money you have in an account. You pay interest when you pay extra money for money you borrow. Banks figure out interest by using equations. Other activities use equations, too. These include saving money, buying a house, and owning a business.

You Try It

A box has a volume = 24 cm^3. It has a length of 3 cm, and a width of 2 cm. Draw a picture of the box. Show the measurements.

- Which value is missing? Identify a variable for the value.

- Write an equation for the situation by substituting values into volume = length × width × height.

- What is the value of the missing height?

The Equations Keep Multiplying

In this image, how many individual eyes do you see? Could you recognize a quick and efficient way to solve this problem, one that involved the use of multiplication? Rather than counting each individual eye one by one, we could multiply the number of rows by the amount of columns. We instantaneously see that 5 columns × 4 rows = 20 eyes.

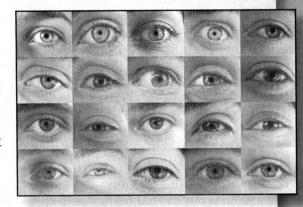

Basic Facts

We say that this statement, *5 × 4 = 20,* is an equation. It is a mathematical statement in which the values on each side must be equal.

We know *5 × 4 = 20* is a mathematical statement. The variable *k* represents an unknown value. It is advisable not to use × for multiplication when variables are involved because of the confusion that this can cause. Instead, we write *5 • k = 20*. When you have uncovered the value for the variable that makes the equation true, you have solved the equation. In this case, you obviously discover the value of *k* as being equal to 4.

Equations with Multiplication

Imagine you have been asked to construct a rectangular soccer field. Despite not having all the dimensions, you know that the total area of the field is 6,000 square feet, and the length is 100 feet, but would you be able to determine the width?

We can use an equation to represent this dilemma. We know that the soccer field is a rectangle and we also know that multiplying the length by the width of a rectangle is the formula for finding area.

So, in words we know that *length × width = area.*

| Assign a variable. | Let w = width |
| Write an equation. | $100 \cdot w = 6{,}000$ |

Now let's find the value of *w*.

| Use fact families. | $6{,}000 \div 100 = w$ |
| Or, guess and check. | $60 = w$ |

It is recommended that you always verify your answer.

Write the original equation.	$100 \cdot w = 6{,}000$
Substitute 60 for *w*. Solve.	$100 \cdot \mathbf{60} = 6{,}000$
So, the sides are balanced!	$6{,}000 = 6{,}000$

The width is 60 feet.

Equations and Formulas

Formulas reveal the relationships between values.

The formula for the volume of this figure is:

volume = length × width × height

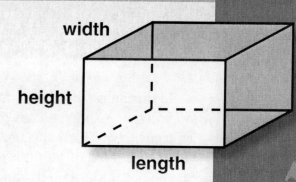

The best place to start is by inserting the values that are known into the formula, and then deducing or solving to find what's missing.

Example: Find the width. The volume is = 40 cm³. The length = 2 cm. The height = 4 cm.

Decide what you must find.	Let w = width
Write the formula.	volume = length × width × height
Put in what you know.	40 cm³ = 2 cm • w • 4 cm
The 8 cm comes from 2 cm × 4 cm.	40 cm³ = 8 cm • w
Use fact families. Or, guess and check.	40 cm³ ÷ 8 cm = w
The width is 5 cm.	5 = w

It is recommended that you always verify your answer.

Write the formula.	volume = length × width × height
Put in what you know.	40 = 2 • 5 • 4
The sides do balance!	40 = 40
The width is 5 cm.	

Multiplication Equations in Our Daily Lives

Ever hear of someone depositing money into a bank and then ignoring it? Then they were surprised years later to find they had accumulated more money. Banks pay **interest**, which means that depositors are awarded a fee for the money they keep in their accounts. Individuals can also owe interest, as when they accrue fees for money that is borrowed. Banks determine the amount of interest by using equations. There are a variety of ways businesses employ the use of formulas and equations. Practically any situation involving money will undoubtedly require formulas, including saving money, acquiring a house, and breaking ground with a new business.

You Try It

A box has a volume = 24 cm³, a length of 3 cm, and a width of 2 cm. Draw a visual representation of the box to help indicate the known measurements.

- Which value is missing? Assign a variable for the value.
- Write an equation by substituting values into volume = length × width × height.
- Determine the value that is missing: in this case, the height.

Equation Writing

Two friends are frosting cupcakes. They need 64. They are having a bake sale. They each frost the same amount. How many do they each frost?

Basic Facts

What did you find? Did you find they each frosted 32 cupcakes? Think of how you found this.

Equation $64 \div 2 = 32$	The $64 \div 2 = 32$ is an **equation**. The $64 \div 2$ is on the left of the equal sign. It has the same value as the 32 on the right.
Equation $64 \div f = 32$ **Variable** f **Solution** $f = 2$	The $64 \div f = 32$ is an **equation**. The **variable** is f. The equation is true when $f = 2$. So $f = 2$ is the **solution** to the equation.

Fact Families

Let's look more at $64 \div 2 = 32$ and at $64 \div f = 32$. Look at the fact families.

$64 \div 2 = 32$	$64 \div 32 = 2$	$64 \div f = 32$	$64 \div 32 = f$
$32 \cdot 2 = 64$	$2 \cdot 32 = 64$	$32 \cdot f = 64$	$f \cdot 32 = 64$

You want to find f in $64 \div f = 32$. Which fact would most help you? The equation $64 \div 32 = f$ would help. This is because $64 \div 32$ tells you that $f = 2$.

Equations with Division

A group of friends frost 64 cupcakes. Each person frosts 16. How many friends helped? We can write an equation. It shows how many friends helped.

number of cupcakes ÷ number of friends = cupcakes frosted by each person

Assign a variable.
It stands for what is missing.
We can let n = number of friends

Write an equation.
$64 \div n = 16$

Find the value of n.
Use fact families. Or, guess and check.
$64 \div 16 = n$
$n = 4$

Check your answer.
Write the original equation
$64 \div n = 16$
Put in 4 for n. Then solve.
$64 \div 4 = 16$
So, the sides are balanced!
$16 = 16$

4 friends frosted the cupcakes.

Writing Equations

Look at some sample equations. Some contain variables. Some do not.

If you have 8 wheels then you have enough wheels for 2 skateboards.
Think: *The 8 wheels must be divided evenly between the 2 boards.*
$$8 \div 4 = 2$$

If you have 40 wheels, then you have enough wheels for 10 skateboards.
Think: *The 40 wheels must be divided evenly among the 10 boards.*
$$40 \div 4 = 10$$

Each equation above is in this form:
number of wheels ÷ 4 wheels per board = number of boards.

Write an equation to show how many wheels you need for 22 skateboards.
Let w = number of wheels
$$w \div 4 = 22$$

Solve the equation to show how many wheels you need for 22 skateboards. This is true when $w = 88$ because $88 \div 4 = 22$. So, 88 wheels are needed for 22 skateboards.

Division Equations in Our Daily Lives

Liquid flows. It does this all the time. That is unless it is stopped. What if there is too much pressure? What if this pushes on what is stopping it? Then it will break through. It will flow again. Equations help us understand. They let us see all the factors. They help us work with everything. They show us this process.

This is useful. It can help in work with dams. Dams are barriers. They stop the flow of water. They control it. Dams give us a source for drinking water. They can make electricity. Dams help stop floods. And they make areas for water fun. They make places to swim. And they make places to boat.

You Try It

A local soccer club has 72 players who will be divided into teams.

- Write an equation that shows the 72 players divided into six teams, with each team having 12 players.

- Think about the information given in the situation above, then explain what n would represent in the equation $72 \div n = 9$.

- What is the value of n in the equation $72 \div n = 9$?

Equation Writing

Two friends are frosting cupcakes. They need 64 for a bake sale. They each frost the same amount. How many do they each frost?

Basic Facts

Did you find that each friend frosted 32 cupcakes? Think of how you found this.

Equation $64 \div 2 = 32$	The statement $64 \div 2 = 32$ is an **equation**. The $64 \div 2$ is on the left of the equal sign. It has the same value as the 32 on the right.
Equation $64 \div f = 32$ **Variable** f **Solution** $f = 2$	The statement $64 \div f = 32$ is an **equation**. The **variable** is f. The equation is true when $f = 2$. So, $f = 2$ is the **solution** to the equation.

Fact Families

Let's look more at $64 \div 2 = 32$ and at $64 \div f = 32$. Compare the fact families.

$64 \div 2 = 32$	$64 \div 32 = 2$	$64 \div f = 32$	$64 \div 32 = f$
$32 \cdot 2 = 64$	$2 \cdot 32 = 64$	$32 \cdot f = 64$	$f \cdot 32 = 64$

Which fact would most help you find f in $64 \div f = 32$. The equation $64 \div 32 = f$ would help. This is because the calculation $64 \div 32$ tells you that $f = 2$.

Equations with Division

Think of friends frosting 64 cupcakes. Each person frosts 16. We can write an equation. It will show how many friends are doing the frosting.

number of cupcakes ÷ number of friends = cupcakes frosted by each person

Assign a variable. It stands for what is missing.	We can let n = number of friends
Write an equation.	$64 \div n = 16$
Now let's find the value of n.	
Use fact families. Or, guess and check.	$64 \div 16 = n$ $n = 4$
You should check your answer.	
Write the original equation	$64 \div n = 16$
Put in 4 for n. Then solve.	$64 \div \mathbf{4} = 16$
So, the sides are balanced!	$16 = 16$

4 friends frosted the cupcakes.

Writing Equations

Let's look at some sample equations. Some have variables. Some do not.

You have 8 wheels. You know you have enough wheels for 2 skateboards.
Think: *The 8 wheels must be divided evenly among the 2 boards.*
$$8 \div 4 = 2$$

You have 40 wheels. You have enough wheels for 10 skateboards.
Think: *The 40 wheels must be divided evenly among the 10 boards.*
$$40 \div 4 = 10$$

Each equation above is in this form.
number of wheels ÷ 4 wheels per board = number of boards

Write an equation. Make it show how many wheels you need for 22 skateboards.
Let w = number of wheels
$$w \div 4 = 22$$

Solve the equation to show how many wheels you need for 22 skateboards. This is true when $w = 88$ because $88 \div 4 = 22$. So, 88 wheels are needed.

Division Equations in Our Daily Lives

Liquid will flow. That is, unless it is stopped. What if there is too much pressure on what is stopping it? Then the liquid will break through. It will flow again. Equations help us understand all the factors. They help us work with everything in this process.

This is useful in work with dams. Dams are barriers. They stop the flow of water. Or they control it. Dams give us a source for drinking water. They can help make electricity. Dams help prevent floods. And they give areas for water recreation. They make places to swim. And they make places to boat.

You Try It

A local soccer club has 72 players who will be divided into teams.

- Write an equation that shows the 72 players divided into six teams, with each team having 12 players.

- Think about the information given in the situation above.
  Explain what n would represent in the equation $72 \div n = 9$.

- What is the value of n in the equation $72 \div n = 9$?

Equation Writing

Two friends are frosting 64 cupcakes for a bake sale. How many cupcakes does each friend frost?

Basic Facts

Did you find that each friend frosted 32 cupcakes? Think of how you found this value.

Equation $64 \div 2 = 32$	The statement $64 \div 2 = 32$ is an **equation**. The $64 \div 2$ on the left of the equal sign has the same value as the 32 on the right of the equal sign.
Equation $64 \div f = 32$ **Variable** f **Solution** $f = 2$	The statement $64 \div f = 32$ is also an **equation**. The **variable** is f. The equation is true when $f = 2$, so $f = 2$ is the **solution** to the equation.

Fact Families

Let's look more at $64 \div 2 = 32$ and at $64 \div f = 32$. Compare the fact families.

$64 \div 2 = 32$	$64 \div 32 = 2$	$64 \div f = 32$	$64 \div 32 = f$
$32 \cdot 2 = 64$	$2 \cdot 32 = 64$	$32 \cdot f = 64$	$f \cdot 32 = 64$

Which fact would most help you find f in $64 \div f = 32$? The equation $64 \div 32 = f$ would help because the calculation $64 \div 32$ tells you that $f = 2$.

Equations with Division

A group of friends are frosting 64 cupcakes. Each person frosts 16 cupcakes. We can write an equation to show how many friends are frosting the 64 cupcakes.

number of cupcakes ÷ number of friends = cupcakes frosted by each person

Assign a variable. It stands for what is missing.	We can let n = number of friends
Write an equation. Now let's find the value of n. Use fact families. Or, guess and check.	$64 \div n = 16$ $64 \div 16 = n$ $n = 4$
You should check your answer. Write the original equation Put in 4 for n. Then solve. So, the sides are balanced!	$64 \div n = 16$ $64 \div \mathbf{4} = 16$ $16 = 16$

4 friends frosted the cupcakes.

Writing Equations

Let's look at some sample equations, both with and without variables.

If you have 8 wheels then you have enough wheels for 2 skateboards.
Think: *The 8 wheels must be divided evenly among the 2 boards.*
$$8 \div 4 = 2$$

If you have 40 wheels, then you have enough wheels for 10 skateboards.
Think: *The 40 wheels must be divided evenly among the 10 boards.*
$$40 \div 4 = 10$$

Each equation above is in this form:
number of wheels ÷ 4 wheels per board = number of boards

Write an equation to show how many wheels you need for 22 skateboards.
Let w = number of wheels
$$w \div 4 = 22$$

Solve the equation to show how many wheels you need for 22 skateboards. This is true when $w = 88$ because $88 \div 4 = 22$. So, 88 wheels are needed.

Division Equations in Our Daily Lives

Liquid will flow unless it is stopped. If there is too much pressure on whatever is holding or stopping it then the liquid will break through and flow again. Equations help us understand and work with all the factors involved in this process.

This is useful in constructing dams. Dams are barriers that stop or control the flow of water. Dams give us a source for drinking water and can help generate electricity. Dams help prevent floods while also giving areas for water recreation such as swimming and boating.

You Try It

A local soccer club has 72 players. These players will be divided into teams.

- Write an equation that shows the 72 players divided into six teams, with each team having 12 players.

- Think about the information given in the situation above.
  Explain what n would represent in the equation $72 \div n = 9$.

- What is the value of n in the equation $72 \div n = 9$?

Equation Writing

Two friends are frosting 64 cupcakes for a bake sale. Assuming that each friend frosts an equal quantity, how many treats does each friend frost?

Basic Facts

You should have determined that each friend frosted 32 cupcakes. Analyze your thought process.

Equation $64 \div 2 = 32$	The statement $64 \div 2 = 32$ is an **equation**. The $64 \div 2$ on the left of the equal sign has the same value as the 32 on the right of the equal sign.
Equation $64 \div f = 32$ **Variable** f **Solution** $f = 2$	The statement $64 \div f = 32$ is also an **equation** that includes the **variable** f. The equation is true when $f = 2$, so $f = 2$ is considered to be the **solution** to the equation.

Fact Families

Let's investigate the equations $64 \div 2 = 32$ and $64 \div f = 32$ more closely. Compare the fact families.

$64 \div 2 = 32$	$64 \div 32 = 2$	$64 \div f = 32$	$64 \div 32 = f$
$32 \cdot 2 = 64$	$2 \cdot 32 = 64$	$32 \cdot f = 64$	$f \cdot 32 = 64$

Which fact would be most relevant for finding the value of f in $64 \div f = 32$? The equation $64 \div 32 = f$ would help because the calculation $64 \div 32$ indicates that $f = 2$.

Equations with Division

A group of friends are frosting 64 cupcakes, with each person frosting 16 cupcakes. Write an equation to represent how many friends are frosting the 64 cupcakes.

number of cupcakes ÷ number of friends = cupcakes frosted by each person

Assign a variable. The variable stands for what is missing.	We can let n = number of friends
Write an equation.	$64 \div n = 16$
Now let's find the value of n. Use fact families. Or, guess and check.	$64 \div 16 = n$ $n = 4$
You should verify your answer. Write the original equation Put in 4 for n. Then solve. So, the sides are balanced!	$64 \div n = 16$ $64 \div \mathbf{4} = 16$ $16 = 16$

4 friends frosted the cupcakes.

Writing Equations

Let's scrutinize some sample equations, both with and without variables.

If you have 8 wheels then you have enough wheels for 2 skateboards.
Think: *The 8 wheels must be divided evenly among the 2 boards.*
$$8 \div 4 = 2$$

If you have 40 wheels, then you have enough wheels for 10 skateboards.
Think: *The 40 wheels must be divided evenly among the 10 boards.*
$$40 \div 4 = 10$$

Each equation above is in this form:
number of wheels ÷ 4 wheels per board = number of boards.

Write an equation to show how many wheels you need for 22 skateboards.
Let w = number of wheels
$$w \div 4 = 22$$

Solve the equation to show how many wheels are required for 22 skateboards. This is true when $w = 88$ because $88 \div 4 = 22$. So, 88 wheels are needed.

Division Equations in Our Daily Lives

Liquid will flow continuously, until something comes along to restrict it. If enough pressure is exerted on whatever is holding or stopping the liquid, then the liquid will break through and flow again. Equations offer a way to interpret and work with all the factors involved in this process.

Think how useful this could be for constructing dams. Dams are barriers that stop or control the flow of water; additionally, they provide a source for drinking water and can help generate electricity. Dams help prevent floods while also giving areas for water recreation such as swimming and boating.

You Try It

A local soccer club has 72 players, and wants to divide these players into teams.

- Write an equation that shows the 72 players divided into six teams, with each team having 12 players.

- Consider the information revealed in the situation above.
  Explain what n would represent in the equation $72 \div n = 9$.

- What is the value of n in the equation $72 \div n = 9$?

Everything Has a Place

What is the biggest penguin? It is the Emperor Penguin! They dive deep. This lets them catch food. They can stay down for 18 minutes. They can swim more than 700 feet down. That is deep!

Think of two penguins. They are looking for food. The first swims down 652 feet. The second dives to 628 feet. Which one swims deeper? How can we tell?

Basic Facts

inequality	This is a mathematical statement. It compares two numbers. Or, it does this for two expressions. Which value is greater? It shows if the left side is. Or, it shows if the right side is.
>	This symbol is *greater than*. It shows that one thing is bigger than something else. Greater means "bigger." Which bird went deeper? The one that went 652 feet did. 652 > 628. 652 is greater than 628.
<	This symbol is *less than*. It shows that one thing is less than something else. Which bird swam less deep? The one that went 628 feet down did. 628 < 652. 628 is less than 652.

Base Ten Blocks

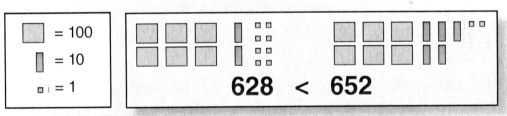

Base ten blocks can help. They can show us that 628 < 652. How? There are fewer blocks in 628. There are more blocks in 652. This is how we can tell.

Place Value and Expanded Form

Look at the **expanded forms**. Both forms show 600. But 20 is less than 50. So 628 is less than 652. Look at the **digits**. Look at the tens place of 6<u>2</u>8. It has a 2. Look at the tens place of 6<u>5</u>2. It has a 5. 2 is less than 5. It does not matter how you look at it. 628 is less than 652.

Number	Base Ten Blocks in Words	Expanded Form
628	6 *hundreds* + 2 *tens* + 8 *units*	600 + 20 + 8
652	6 *hundreds* + 5 *tens* + 2 *units*	600 + 50 + 2

Multiple Methods, Multiple Numbers

Number	Base Ten Blocks in Words	Expanded Form
2,935	2 *thousands* + 9 *hundreds* + 3 *tens* + 5 *units*	2,000 + 900 + 30 + 5
3,150	3 *thousands* + 1 *hundred* + 5 *tens* + 0 *units*	3,000 + 100 + 50 + 0
2,873	2 *thousands* + 8 *hundreds* + 7 *tens* + 3 *units*	2,000 + 800 + 70 + 3
2,981	2 *thousands* + 9 *hundreds* + 8 *tens* + 1 *unit*	2,000 + 900 + 80 + 1

Look at the numbers below from the chart. They are in order. They start from least. They go to greatest. How do you put them in order? What should you do first? Look at the largest digits! First look at the thousands. Compare them. Then move down. Go to the hundreds. Then go to the tens. Then look at the units. Only go as far as you need.

$$2{,}873 \quad 2{,}935 \quad 2{,}981 \quad 3{,}150$$

The numbers below are in order. They start at greatest. They go to least. Here you use the same steps as before. You start with the thousands. Then move down. You end with the ones. This time you just put the numbers in a different order.

$$3{,}150 \quad 2{,}981 \quad 2{,}935 \quad 2{,}873$$

Inequalities in Our Daily Lives

Computers run based on their instructions. These are given in their programs. One instruction is known as "if—then."

Think about playing a video game. You can keep the points from old games. The points from your new game are added in. The game checks your total points. If you have enough you get to go to the next level. Your game might be programmed like this:

IF the old points + the new points > 100,000, **THEN** start the next level.

You Try It

Look at these numbers: 28, 621, 562, and 617.

- Show the numbers on a number line.
- Write the numbers. Put them in order from least to greatest.
- Explain whether 621 < 617 or 621 > 617.

Everything Has a Place

What is the biggest penguin? It is the Emperor Penguin! They dive underwater. This lets them catch their food. They can stay down for 18 minutes. They can swim more than 700 feet below the surface!

Think of two penguins. They are racing for food. The first swims down 652 feet. The second dives down 628 feet. Which one swims to the greater depth? How can we tell?

Basic Facts

inequality	This is a mathematical statement. It compares two numbers or two expressions. It shows if the value of the left side is greater or less than that on the right side.
>	This sign is *greater than*. This is a mathematical symbol. It shows that something is bigger than something else. Greater means "bigger." Which penguin went deeper? The one with a depth of 652 feet swam deeper. 652 > 628. 652 is greater than 628.
<	This sign is *less than*. This is a mathematical symbol. It shows that something is less than something else. Which penguin swam less deep? The one that went 628 feet down did not swim as deep. 628 < 652. 628 is less than 652.

Base Ten Blocks

Base ten blocks can show us that 628 < 652. There are fewer in 628 than in 652.

Place Value and Expanded Form

Look at the **expanded forms**. Both forms show 600. But 20 is less than 50. So, 628 is less than 652. Look at the **digits**. Look at the 2 in the tens place of 6<u>2</u>8. It is less than the 5 in the tens place of 6<u>5</u>2. It does not matter how you look at it. 628 is less than 652.

Number	Base Ten Blocks in Words	Expanded Form
628	6 *hundreds* + 2 *tens* + 8 *units*	600 + 20 + 8
652	6 *hundreds* + 5 *tens* + 2 *units*	600 + 50 + 2

Multiple Methods, Multiple Numbers

Number	Base Ten Blocks in Words	Expanded Form
2,935	2 thousands + 9 hundreds + 3 tens + 5 units	2,000 + 900 + 30 + 5
3,150	3 thousands + 1 hundred + 5 tens + 0 units	3,000 + 100 + 50 + 0
2,873	2 thousands + 8 hundreds + 7 tens + 3 units	2,000 + 800 + 70 + 3
2,981	2 thousands + 9 hundreds + 8 tens + 1 unit	2,000 + 900 + 80 + 1

The numbers from the chart are shown in order. They go from least to greatest. How do you put them in order? Compare the largest digits first. Then move down. In this case, first look at the thousands. Then go to the hundreds. Then go to the tens. Finally, if needed, look at the units.

$$2,873 \quad 2,935 \quad 2,981 \quad 3,150$$

The numbers below are ordered. They go from greatest to least. Here you do the same thing. You start at the thousands. You go down to the ones. But the end number order is reversed.

$$3,150 \quad 2,981 \quad 2,935 \quad 2,873$$

Inequalities in Our Daily Lives

Computers run based on their instructions. These are given in their programs. One type of instruction is known as "if—then."

Think about playing a video game. You can keep the points from old games. The points from your new game are added to your old points. If you earn enough total points, then you get to move on to the next level. Your game might be programmed like this:

IF the old points + the new points > 100,000, **THEN** start the next level.

You Try It

Look at these numbers: 28, 621, 562, and 617.

- Show the numbers on a number line.
- Write the numbers. Put them in order from least to greatest.
- Use base ten blocks in words, or use expanded form to explain whether 621 < 617 or 621 > 617.

Everything Has a Place

Did you know that the largest penguin is the Emperor Penguin? They dive underwater to catch their food and can stay down for up to 18 minutes. They have been known to swim more than 700 feet below the surface of the water!

Imagine two Emperor penguins racing for food. The first swims down 652 feet. The second dives down 628 feet. Which penguin swims to the greatest depth?

Basic Facts

inequality	This is a mathematical statement. It compares two numbers or expressions. An inequality shows if values on the left side are greater or less than values on the right side.
>	This symbol is *greater than*. This is a mathematical symbol to show that something is larger or greater than something else. The first penguin, with a depth of 652 feet, swims the deepest. 652 > 628. 652 is greater than 628.
<	This symbol is *less than*. This is a mathematical symbol to show that something is smaller, or less than something else. The second penguin, with a depth of 628 feet, does not swim as deep. 628 < 652. 628 is less than 652.

Base Ten Blocks

Base ten blocks can show us that 628 < 652. There are fewer in 628 than in 652.

Place Value and Expanded Form

Look at the **expanded forms**. Both forms show 600, but 20 is less than 50, so 628 is less than 652. Look at the **digits**. The 2 in the tens place of 6<u>2</u>8 is less than the 5 in the tens place of 6<u>5</u>2. Either way you look at it, 628 is less than 652.

Number	Base Ten Blocks in Words	Expanded Form
628	6 *hundreds* + 2 *tens* + 8 *units*	600 + 20 + 8
652	6 *hundreds* + 5 *tens* + 2 *units*	600 + 50 + 2

Multiple Methods, Multiple Numbers

Number	Base Ten Blocks in Words	Expanded Form
2,935	2 *thousands* + 9 *hundreds* + 3 *tens* + 5 *units*	2,000 + 900 + 30 + 5
3,150	3 *thousands* + 1 *hundred* + 5 *tens* + 0 *units*	3,000 + 100 + 50 + 0
2,873	2 *thousands* + 8 *hundreds* + 7 *tens* + 3 *units*	2,000 + 800 + 70 + 3
2,981	2 *thousands* + 9 *hundreds* + 8 *tens* + 1 *unit*	2,000 + 900 + 80 + 1

The numbers from the chart are shown in order from least to greatest. To put them in order, compare the largest digits first, then move down. In this case, first compare the thousands, then the hundreds, then the tens. Finally, if needed, compare the units.

$$2,873 \quad 2,935 \quad 2,981 \quad 3,150$$

The numbers below are ordered from greatest to least. The process is the same, but the order is reversed.

$$3,150 \quad 2,981 \quad 2,935 \quad 2,873$$

Inequalities in Our Daily Lives

Computers run based on the instructions that are given in their programs. One type of instruction is known as "if—then."

Imagine you are playing a video game. You are able to keep the points from old games. The points you earn from your new game are added to your old points. If you earn enough total points, then you get to move on to the next level. Your game might be programmed similar to:

IF the old points + the new points > 100,000, **THEN** start the next level.

You Try It

Look at these numbers: 28, 621, 562, and 617.

- Show the numbers on a number line.
- Write the numbers in order from least to greatest.
- Use base ten blocks in words, or use expanded form to explain whether 621 < 617 or 621 > 617.

Everything Has a Place

Did you know the largest penguin is the Emperor Penguin? They dive deep underwater to catch their food and can stay down for 18 minutes. They have been known to swim more than 700 feet below the surface of the water!

Imagine two Emperor penguins racing for food. The first swims down 652 feet under the surface, while the second dives down 628 feet. Which penguin swims to a greater depth?

Basic Facts

inequality	This is a mathematical statement that compares two numbers or expressions to show if the value of the left side is greater or less than the value of the right side.
>	*Greater than* is a mathematical symbol to show that something is larger or greater than something else. The first penguin, with a depth of 652 feet, swims the deepest. 652 > 628. 652 is greater than 628.
<	*Less than* is a mathematical symbol to show that something is smaller or less than something else. The second penguin, with a depth of 628 feet, does not swim as deep. 628 < 652. 628 is less than 652.

Base Ten Blocks

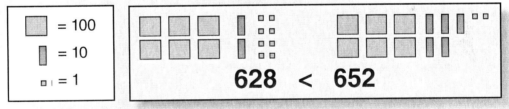

Base ten blocks can show us that 628 < 652. There are fewer in 628 than in 652.

Place Value and Expanded Form

Look at the **expanded forms** of these numbers. Both forms show 600, but 20 is less than 50, so 628 is less than 652. Look at the **digits** in these numbers. The 2 in the tens place of 6<u>2</u>8 is less than the 5 in the tens place of 6<u>5</u>2. Either way you look at it, 628 is less than 652.

Number	Base Ten Blocks in Words	Expanded Form
628	6 *hundreds* + 2 *tens* + 8 *units*	600 + 20 + 8
652	6 *hundreds* + 5 *tens* + 2 *units*	600 + 50 + 2

Multiple Methods, Multiple Numbers

Number	Base Ten Blocks in Words	Expanded Form
2,935	2 thousands + 9 hundreds + 3 tens + 5 units	2,000 + 900 + 30 + 5
3,150	3 thousands + 1 hundred + 5 tens + 0 units	3,000 + 100 + 50 + 0
2,873	2 thousands + 8 hundreds + 7 tens + 3 units	2,000 + 800 + 70 + 3
2,981	2 thousands + 9 hundreds + 8 tens + 1 unit	2,000 + 900 + 80 + 1

The numbers from the chart are shown below in order from least to greatest. To put them in order, compare the largest digits first then move down. In this case, first compare the thousands, then the hundreds, then the tens, and finally, if needed, compare the units.

2,873   2,935   2,981   3,150

The numbers below are ordered from greatest to least. The process is the same as above, but the order is reversed.

3,150   2,981   2,935   2,873

Inequalities in Our Daily Lives

Computers function based on the instructions they are given by their programs. One type of instruction is known as "if—then."

Imagine you are playing a video game in which you are able to retain the points from old games. The points you earn from your new game are consolidated with your old points and if you accumulate enough total points then you get to move to the next level. Your game might be programmed similar to:

IF the old points + the new points > 100,000, **THEN** commence the next level.

You Try It

Consider the following numbers: 28, 621, 562, and 617.

- Show the numbers on a number line.
- Write the numbers in order from least to greatest.
- Use base ten blocks in words, or use expanded form to explain whether 621 < 617 or 621 > 617.

Moving Around

Here are some actions. Which one is commutative? How do you know?

1. Wash the dishes. Then dry them with a towel.
2. Put on your pants. Then put on your shirt.

Basic Facts

See these examples. Think about each operation. Are the answers the same for both problems? Are they the same for the subtraction? Are they the same for the multiplication? Are they the same for division?

Addition	Subtraction
$8 + 2 = 10$ $2 + 8 = 10$	$10 - 2 = 8$ $2 - 10 =$ a value less than zero
Multiplication	**Division**
$4 \times 2 = 8$ $2 \times 4 = 8$	$4 \div 2 = 2$ $2 \div 4 = \frac{1}{2}$ The two circles would each have to be cut in half to make 4 groups.

Commutative Property for Addition

What did you find? Were answers for both addition problems the same?

We saw that $8 + 2 = 10$ and $2 + 8 = 10$, so $8 + 2 = 2 + 8$.

We can say: $a + b = b + a$

We can add in any order. It is our choice. Any number can come first.

Basic Facts (cont.)

Commutative Property for Multiplication

What did you find? Were both multiplication answers the same?

We saw that $4 \times 2 = 8$ and $2 \times 4 = 8$, so $4 \times 2 = 2 \times 4$.

In general we can say: $a \times b = b \times a$.

When we multiply numbers, we can also do so in any order that we choose.

Filling in the Blank

The commutative property is extremely useful. We can find missing values. Look at each example. Look at the matching answer. In each case, the left-hand side does equal the right-hand side.

124 + 52 = __ + 124	21 × 40 = 40 × __
124 + 52 = <u>52</u> + 124	21 × 40 = 40 × <u>21</u>
176 = 176	840 = 840

Adding Many Numbers

We can add in any order. We can find easy ways to add a string of numbers.

7 + 12 + 3 + 5 + 8 =	Switch the 12 and 3. Switch the 5 and 8.
7 + 3 + 12 + 8 + 5	Remember that order does not matter.
10 + 20 + 5	We can move the numbers into whatever position is convenient, as needed.

Moving the numbers around helps. We find easy groups of 10. Multiples of 10 make the addition easier.

Commutative Property in Our Daily Lives

Do you complete your math homework first? Then you start your spelling homework? Or, you might complete spelling first. You do math later. Even without numbers, there are examples where the order does not matter. This is also the commutative property.

You Try It

What is the commutative property? Use $6 \times 8 = \underline{} \times 6$ as you explain.

Use the commutative property to add 7 + 4 + 2 + 13 + 6. Show which pairs of numbers you add first.

Moving Around

Which action below is commutative and how do you know?

1. Wash the dishes. Then dry them with a towel.
2. Put on your pants. Then put on your shirt.

Basic Facts

Look at each operation in these examples. Are the answers for both addition problems the same? Are they the same for the subtraction problems and for the multiplication problems? Are they the same for division?

Addition	Subtraction
$8 + 2 = 10$ $2 + 8 = 10$	$10 - 2 = 8$ $2 - 10 =$ a value less than zero
Multiplication	**Division**
$4 \times 2 = 8$ $2 \times 4 = 8$	$4 \div 2 = 2$ $2 \div 4 = \frac{1}{2}$ The two circles would each have to be cut in half to make 4 groups.

Commutative Property for Addition

The answer for both addition problems was the same.

We saw that $8 + 2 = 10$ and $2 + 8 = 10$, so $8 + 2 = 2 + 8$.

In general, we can say: $a + b = b + a$.

We can add numbers. We can add them in any order that we choose.

Basic Facts (cont.)

Commutative Property for Multiplication

Did you notice that the answer for both multiplication problems were the same?

We saw that $4 \times 2 = 8$ and $2 \times 4 = 8$, so $4 \times 2 = 2 \times 4$.

In general we can say: $a \times b = b \times a$.

When we multiply numbers, we can do so in any order that we choose.

Filling in the Blank

The commutative property can help us find missing values. Look at each example and each matching answer. In each example below, the last line shows that the left-hand side *does* equal the right-hand side.

$124 + 52 = \_\_ + 124$ $21 \times 40 = 40 \times \_\_$

$124 + 52 = \underline{52} + 124$ $21 \times 40 = 40 \times \underline{21}$

$176 = 176$ $840 = 840$

Adding Many Numbers

The fact that we can add in any order helps us add a string of numbers.

$7 + 12 + 3 + 5 + 8 =$	Switch the 12 and 3. Switch the 5 and 8.
$7 + 3 + 12 + 8 + 5$	Order does not matter.
$10 + 20 + 5$	We can move the numbers as needed.

Moving the numbers around helps us find easy groups of 10. Multiples of 10 make the addition easier.

Commutative Property in Our Daily Lives

You might complete your math homework and then your spelling homework. Or, you might complete your spelling homework later and then your math homework. Even without numbers, there are examples where the order does not matter. These situations also show the commutative property.

You Try It

What is the commutative property? Use $6 \times 8 = \_\_ \times 6$ as you explain.

Use the commutative property to add $7 + 4 + 2 + 13 + 6$. Show which pairs of numbers you add first.

Moving Around

Read the series of actions below. Can you determine which of them is commutative? How did you come to that decision?

1. Wash the dishes. Then dry them with a towel.
2. Put on your pants. Then put on your shirt.

Basic Facts

Spend a minute studying the problems below. Clarify how the operations work in each problem. How do the answers for the addition problems compare? How do the answers for the subtraction problems compare? What about the the multiplication problems? What about the division?

Addition		Subtraction	
$8 + 2 = 10$	$2 + 8 = 10$	$10 - 2 = 8$	$2 - 10$ = a value less than zero
Multiplication		**Division**	
$4 \times 2 = 8$	$2 \times 4 = 8$	$4 \div 2 = 2$	$2 \div 4 = \frac{1}{2}$
			The two circles would each have to be cut in half to make 4 groups.

Commutative Property for Addition

Did you conclude that the answer for the addition problems was identical?

We saw that $8 + 2 = 10$ and $2 + 8 = 10$, so $8 + 2 = 2 + 8$.

In general, we can say: $a + b = b + a$.

We can add numbers in any order. It is our choice.

Basic Facts (cont.)

Commutative Property for Multiplication

Did you conclude that the answers for both multiplication problems were identical?

We saw that $4 \times 2 = 8$ and $2 \times 4 = 8$, so $4 \times 2 = 2 \times 4$.

We can generalize that: $a \times b = b \times a$.

When we add strings of consecutive numbers, we can do so in any order.

Filling in the Blank

The commutative property is definitely a useful thing to know. It lets us easily and quickly find missing values. As evidenced in each example below, when you look at each matching answer it is evident that the left-hand side equals the right-hand side. The last line shows us:

124 + 52 = __ + 124	21 × 40 = 40 × __
124 + 52 = <u>52</u> + 124	21 × 40 = 40 × <u>21</u>
176 = 176	840 = 840

Adding Many Numbers

The fact that we can add in any order helps us. It makes it easy to add a string of numbers.

7 + 12 + 3 + 5 + 8 =	Switch the 12 and 3. Switch the 5 and 8.
7 + 3 + 12 + 8 + 5	Order does not matter.
10 + 20 + 5	We can move the numbers as needed.

Moving the numbers helps us find easy groups of 10. Multiples of 10 make the addition easier.

Commutative Property in Our Daily Lives

You might finish your math homework. Then you may do your spelling homework next. Or, you might do your spelling homework first, and then you may do your math. You don't need numbers. There are examples where the order does not matter. These situations show the commutative property.

You Try It

What is the commutative property? Use $6 \times 8 = \_\_ \times 6$ as you explain.

Use the commutative property to add 7 + 4 + 2 + 13 + 6. Show which pairs of numbers you add first.

Moving Around

When you examine these sequences of events, can you distinguish which are commutative? How did you come to that conclusion?

1. Wash the dishes. Then dry them with a towel.
2. Put on your pants. Then put on your shirt.

Basic Facts

Observe each operation in the examples below. What do you notice about the solutions for both addition problems? How about for the subtraction problems? Now look at the multiplication and division problems. Can you make any generalizations based on these examples?

Addition	Subtraction
$8 + 2 = 10$ $2 + 8 = 10$	$10 - 2 = 8$ $2 - 10 =$ a value less than zero

Multiplication	Division
$4 \times 2 = 8$ $2 \times 4 = 8$	$4 \div 2 = 2$ $2 \div 4 = \frac{1}{2}$
	The two circles would each have to be cut in half to make 4 groups.

Commutative Property for Addition

Hopefully you recognized that both answers for addition were identical.

We saw that $8 + 2 = 10$ and $2 + 8 = 10$, so $8 + 2 = 2 + 8$.

We can generalize that: $a + b = b + a$.

When adding numbers, the order in which we do so is arbitrary and irrelevant.

Basic Facts (cont.)

Commutative Property for Multiplication

Hopefully you recognized that both multiplication answers were identical.

We saw that $4 \times 2 = 8$ and $2 \times 4 = 8$, so $4 \times 2 = 2 \times 4$.

We can generalize that: $a \times b = b \times a$.

Like addition, we can multiply numbers in any order. The result remains constant.

Filling in the Blank

Knowing the commutative property can be helpful when trying to find missing values. Look at each example and its matching answer below. The last line validates the property, since the results on each side of the equal sign are identical.

124 + 52 = __ + 124	21 × 40 = 40 × __
124 + 52 = <u>52</u> + 124	21 × 40 = 40 × <u>21</u>
176 = 176	840 = 840

Adding Many Numbers

Because we know that order is irrelevant, we can manipulate the numbers to find more efficient methods of adding long strings of numbers.

7 + 12 + 3 + 5 + 8 =	Transpose 12 and 3. Transpose 5 and 8.
7 + 3 + 12 + 8 + 5	The order doesn't matter.
10 + 20 + 5	We can rearrange the numbers as needed.

Manipulating the order of the numbers allows us to create groups of 10s. Adding groups of 10 is easier, even for mental math.

Commutative Property in Our Daily Lives

You have a dilemma: complete your math homework first, and then tackle spelling? Or, dive into spelling first, and save math for later? The commutative property is not limited to numerical situations. A variety of other examples demonstrate that order is not important. These situations show the commutative property.

You Try It

Explain the commutative property using $6 \times 8 =$ __ $\times 6$ as you explain.

Apply the commutative property to add $7 + 4 + 2 + 13 + 6$. Indicate the sets of numbers you added first.

In a Group

Who do you like to associate with? What does that mean? It means who you like being with.

Basic Facts

Look at the examples below. Think about the operations. Are all the answers the same? Are the answers for both addition problems the same? How about for the subtraction problems? Are they the same for the multiplication? Are they are same for division?

Addition		Subtraction	
$8 + 4 + 2$		$8 - 4 - 2$	
$(8 + 4) + 2$	$8 + (4 + 2)$	$(8 - 4) - 2$	$8 - (4 - 2)$
$= 12 + 2$	$= 8 + 6$	$= 4 - 2$	$= 8 - 2$
$= 14$	$= 14$	$= 2$	$= 6$
Multiplication		Division	
$8 \times 4 \times 2$		$8 \div 4 \div 2$	
$(8 \times 4) \times 2$	$8 \times (4 \times 2)$	$(8 \div 4) \div 2$	$8 \div (4 \div 2)$
$= 32 \times 2$	$= 8 \times 8$	$= 2 \div 2$	$= 8 \div 2$
$= 64$	$= 64$	$= 1$	$= 4$

Associative Property for Addition

The answers for both addition problems above were the same.

We saw that $(8 + 4) + 2 = 14$ and $8 + (4 + 2) = 14$, so $(8 + 4) + 2 = 8 + (4 + 2)$.

In general, we can say: $(a + b) + c = a + (b + c)$.

The sum stays the same. This is even when the groupings change. Think of the numbers. Think of them as willing to associate with each other. They can be grouped. We can put them in any order that we want.

Associative Property for Multiplication

The answers for both addition problems above were the same.

We saw that $(8 \times 4) \times 2 = 64$ and $8 \times (4 \times 2) = 64$, so $(8 \times 4) \times 2 = 8 \times (4 \times 2)$.

In general, we can say: $(a \times b) \times c = a \times (b \times c)$.

The product stays the same. This is even when the groupings change. Think of the numbers being multiplied. Think of them as willing to **associate** with each other. They are "friends." So they can be grouped. We can put them in any order we want.

Using the Associative Property

The associative property can help us. It lets us find missing values. Look at each example. Look at each answer. The left-hand side does equal the right-hand side. The last line shows that.

(2 + 6) + 4 = 2 + (__ + 4) (2 + 6) + 4 = 2 + (6 + 4) 8 + 4 = 2 + 10 12 = 12	The 6 goes in the blank. Add inside the parentheses first. The two sides are equal.
(3 × 7) × 2 = 3 × (__ × 2) (3 × 7) × 2 = 3 × (7 × 2) 21 × 2 = 3 × 14 42 = 42	The 7 goes in the blank. Work inside the parentheses first. The two sides are equal.

Associative Property Can Make Calculations Easier

Look at the problems below.

4 × (25 × 27)   (313 + 2) + 498

= (4 × 25) × 27   = 313 + (2 + 498)

= 100 × 27   = 313 + 500

= 2,700   = 813

Associative Property in Our Daily Lives

The associative property can be useful. It can help when you are trying to add things. You may use it at the store. You want to buy three items. You can add them in any order. You can look for easy groups. This will help you know the answer. It will let you see if you have enough money.

You Try It

Fill in the blank: 13 + (7 + 4) = (13 + ___) + 4

Multiply. Use the associative property. (23 × 50) × 2

In a Group

Who do you like to associate with? What does that mean? It means who you like being with.

Basic Facts

Look at the examples below. Think about the operations. Are the answers for both addition problems the same? How about for the subtraction problems? Are they the same for the multiplication? Are they the same for division?

Addition		Subtraction	
$8 + 4 + 2$		$8 - 4 - 2$	
$(8 + 4) + 2$	$8 + (4 + 2)$	$(8 - 4) - 2$	$8 - (4 - 2)$
$= 12 + 2$	$= 8 + 6$	$= 4 - 2$	$= 8 - 2$
$= 14$	$= 14$	$= 2$	$= 6$
Multiplication		Division	
$8 \times 4 \times 2$		$8 \div 4 \div 2$	
$(8 \times 4) \times 2$	$8 \times (4 \times 2)$	$(8 \div 4) \div 2$	$8 \div (4 \div 2)$
$= 32 \times 2$	$= 8 \times 8$	$= 2 \div 2$	$= 8 \div 2$
$= 64$	$= 64$	$= 1$	$= 4$

Associative Property for Addition

What did you find? Did you find that the answers for both addition problems above were the same?

We saw that $(8 + 4) + 2 = 14$ and $8 + (4 + 2) = 14$, so $(8 + 4) + 2 = 8 + (4 + 2)$.

In general, we can say: $(a + b) + c = a + (b + c)$.

The sum remains the same. Even when the groupings of addends change. Think of the numbers as associating with each other. They can be grouped in any order we want.

Associative Property for Multiplication

Did you find that the answers for both multiplication problems above were the same?

We saw that $(8 \times 4) \times 2 = 64$ and $8 \times (4 \times 2) = 64$, so $(8 \times 4) \times 2 = 8 \times (4 \times 2)$.

We can generalize that: $(a \times b) \times c = a \times (b \times c)$.

The product remains the same. This is even when the groupings change. Think of the numbers being multiplied as willing to **associate** with each other. They can be grouped in any order we want.

Using the Associative Property

The associative property can help us. It lets us find missing values. Look at each example. Look at each matching answer. The last line shows that the left-hand side does equal the right-hand side.

(2 + 6) + 4 = 2 + (__ + 4) (2 + 6) + 4 = 2 + (6 + 4) 8 + 4 = 2 + 10 12 = 12	The 6 goes in the blank. Add inside the parentheses first. The two sides are equal.
(3 × 7) × 2 = 3 × (__ × 2) (3 × 7) × 2 = 3 × (7 × 2) 21 × 2 = 3 × 14 42 = 42	The 7 goes in the blank. Work inside the parentheses first. The two sides are equal.

Associative Property Can Make Calculations Easier

Look at the problems below.

4 × (25 × 27)　　　　(313 + 2) + 498

= (4 × 25) × 27　　　 = 313 + (2 + 498)

= 100 × 27　　　　　 = 313 + 500

= 2,700　　　　　　　= 813

Associative Property in Our Daily Lives

The associative property can be used when you are trying to add things. You may want to buy three items at the store. You can add them in any order to find the total. This will help you know whether you have enough money to buy them. This is because of the associative property.

You Try It

Fill in the blank: 13 + (7 + 4) = (13 + ___) + 4

Use the associative property to multiply: (23 × 50) × 2

In a Group

Who do you like to associate with? That's just another way to say who do you like being with?

Basic Facts

Observe each operation in the examples below. Are the answers identical for the addition problems? How about the subtraction problems, or the multiplication problems? Are they the same for division?

Addition		Subtraction	
$8 + 4 + 2$		$8 - 4 - 2$	
$(8 + 4) + 2$	$8 + (4 + 2)$	$(8 - 4) - 2$	$8 - (4 - 2)$
$= 12 + 2$	$= 8 + 6$	$= 4 - 2$	$= 8 - 2$
$= 14$	$= 14$	$= 2$	$= 6$
Multiplication		Division	
$8 \times 4 \times 2$		$8 \div 4 \div 2$	
$(8 \times 4) \times 2$	$8 \times (4 \times 2)$	$(8 \div 4) \div 2$	$8 \div (4 \div 2)$
$= 32 \times 2$	$= 8 \times 8$	$= 2 \div 2$	$= 8 \div 2$
$= 64$	$= 64$	$= 1$	$= 4$

Associative Property for Addition

You should recognize that the results for the addition problems above were identical.

We saw that $(8 + 4) + 2 = 14$ and $8 + (4 + 2) = 14$, so $(8 + 4) + 2 = 8 + (4 + 2)$.

We can generalize that: $(a + b) + c = a + (b + c)$.

The sum remains the same even when the groupings of addends change. Think of the numbers being added as willing to associate with each other. They are "friends," so they can be grouped in any order that we desire.

Associative Property for Multiplication

You should recognize that the answers for both multiplication problems were identical.

We saw that $(8 \times 4) \times 2 = 64$ and $8 \times (4 \times 2) = 64$, so $(8 \times 4) \times 2 = 8 \times (4 \times 2)$.

We can generalize that: $(a \times b) \times c = a \times (b \times c)$.

The product remains the same even when the groupings of factors change. Consider the numbers being multiplied as willing to **associate** with each other. They are "friends," so they can be grouped in any order that suits us.

Using the Associative Property

The associative property can help us find missing values. Look at each example and each matching answer. In each, the last line shows that the left-hand side *does* equal the right-hand side.

(2 + 6) + 4 = 2 + (__ + 4) (2 + 6) + 4 = 2 + (6 + 4) 8 + 4 = 2 + 10 12 = 12	The 6 goes in the blank. Add inside the parentheses first. The two sides are equal.
(3 × 7) × 2 = 3 × (__ × 2) (3 × 7) × 2 = 3 × (7 × 2) 21 × 2 = 3 × 14 42 = 42	The 7 goes in the blank. Work inside the parentheses first. The two sides are equal.

Associative Property Can Make Calculations Easier

Look at the problems below.

4 × (25 × 27)   (313 + 2) + 498

= (4 × 25) × 27   = 313 + (2 + 498)

= 100 × 27   = 313 + 500

= 2,700   = 813

Associative Property in Our Daily Lives

The associative property can be used when you are trying to add things together. For example, you may want to buy three items at the store. Because of the associative property, you can add them in any order to find the total, which will help you know whether you have enough money to buy them.

You Try It

Fill in the blank: 13 + (7 + 4) = (13 + ___) + 4

Use the associative property to multiply: (23 × 50) × 2

In a Group

Are there people with whom you like to associate? In other words, are there individuals you prefer to be with?

Basic Facts

Consider each operation in the examples below. What do you notice about the addition problems? How about the subtraction problems? Anything stand out about the multiplication problems? What about for division?

Addition		Subtraction	
$8 + 4 + 2$		$8 - 4 - 2$	
$(8 + 4) + 2$	$8 + (4 + 2)$	$(8 - 4) - 2$	$8 - (4 - 2)$
$= 12 + 2$	$= 8 + 6$	$= 4 - 2$	$= 8 - 2$
$= 14$	$= 14$	$= 2$	$= 6$
Multiplication		**Division**	
$8 \times 4 \times 2$		$8 \div 4 \div 2$	
$(8 \times 4) \times 2$	$8 \times (4 \times 2)$	$(8 \div 4) \div 2$	$8 \div (4 \div 2)$
$= 32 \times 2$	$= 8 \times 8$	$= 2 \div 2$	$= 8 \div 2$
$= 64$	$= 64$	$= 1$	$= 4$

Associative Property for Addition

Did you recognize that the results of both addition problems above were the same?

We saw that $(8 + 4) + 2 = 14$ and $8 + (4 + 2) = 14$, so $(8 + 4) + 2 = 8 + (4 + 2)$.

We can generalize that: $(a + b) + c = a + (b + c)$.

The sum remains the same even when the groupings of addends change. Think of the numbers being added as willing to associate with one another. They can be grouped in any order that we desire. We can make the determination based on which order makes the problem easier to work.

Associative Property for Multiplication

Did you recognize that the results for both multiplication problems above were identical, regardless of the fact that they were grouped differently?

We saw that $(8 \times 4) \times 2 = 64$ and $8 \times (4 \times 2) = 64$, so $(8 \times 4) \times 2 = 8 \times (4 \times 2)$.

We can generalize that: $(a \times b) \times c = a \times (b \times c)$.

The product remains the same despite the fact that the groupings of factors changed. Consider the numbers being multiplied as willing to **associate** with one another. They are "friends," so they can be grouped in any order that suits us. Again, we can make the determination based off of which order will facilitate the easiest computations.

Using the Associative Property

Understanding the associative property can be helpful for finding missing values. Consider each example and each matching answer. Notice that the result validates the concept of the property, since the left-hand side does, in fact, equal the right-hand side.

$(2 + 6) + 4 = 2 + (\_\_ + 4)$ $(2 + 6) + 4 = 2 + (\underline{6} + 4)$ $8 + 4 = 2 + 10$ $12 = 12$	The 6 goes in the blank. Add inside the parentheses first. The two sides are equal.
$(3 \times 7) \times 2 = 3 \times (\_\_ \times 2)$ $(3 \times 7) \times 2 = 3 \times (\underline{7} \times 2)$ $21 \times 2 = 3 \times 14$ $42 = 42$	The 7 goes in the blank. Work inside the parentheses first. The two sides are equal.

Associative Property Can Make Calculations Easier

Observe the problems below.

$4 \times (25 \times 27)$

$= (4 \times 25) \times 27$

$= 100 \times 27$

$= 2,700$

$(313 + 2) + 498$

$= 313 + (2 + 498)$

$= 313 + 500$

$= 813$

Associative Property in Our Daily Lives

Use the associative property when attempting to add things together. For example, you are considering purchasing three items at the store. You aren't entirely certain that you have enough money, but you want to make the math as easy as possible because you are doing it in your head. The associative property indicates you can add the amounts in any order to find the total, so you should be able to know whether you have enough money to buy the items.

You Try It

Fill in the blank in this expression: $13 + (7 + 4) = (13 + \_\_) + 4$

Use the associative property to multiply: $(23 \times 50) \times 2$

References Cited

August, D. and T. Shanahan (Eds). 2006. Developing literacy in second-language learners: Report of the National Literacy Panel on language-minority children and youth. Mahwah, NJ: Lawrence Erlbaum Associates, Inc.

Common Core State Standards Initiative. 2010. *The standards: Language arts.* (Accessed October 2010.) http://www.corestandards.org/the-standards/languagearts.

Marzano, R., D. Pickering, and J. Pollock. 2001. *Classroom instruction that works.* Alexandria, VA: Association for Supervision and Curriculum Development.

Tomlinson, C.A. 2000. *Leadership for Differentiating Schools and Classrooms.* Alexandria, VA: Association for Supervision and Curriculum Development.

Vygotsky, L.S. 1978. *Mind and society: The development of higher mental processes.* Cambridge, MA: Harvard University Press.

Contents of Teacher Resource CD

NCTM Mathematics Standards

The National Council of Teachers of Mathematics (NCTM) standards are listed in the chart on page 20, as well as on the Teacher Resource CD: *nctm.pdf*. TESOL standards are also included: *TESOL.pdf*.

Text Files

The text files include the text for all four levels of each reading passage. For example, the Various Variables text (pages 21–28) is the *various-variables.doc* file.

PDF Files

The full-color PDFs provided are each eight pages long and contain all four levels of a reading passage. For example, the Various Variables text (pages 21–28) is the *various-variables.pdf* file.

Text Title	Text File	PDF
Various Variables	various-variables.doc	various-variables.pdf
Shaping Up	shaping-up.doc	shaping-up.pdf
Sometimes the Change Is Consistent	change-is-consistent.doc	change-is-consistent.pdf
Sometimes the Change Changes	change-changes.doc	change-changes.pdf
It's All Organized	all-organized.doc	all-organized.pdf
Express It Mathematically	express-it.doc	express-it.pdf
Expressing More…Mathematically	expressing-more.doc	expressing-more.pdf
Many Ways to Look at It	ways-to-look-at-it.doc	ways-to-look-at-it.pdf
Adding Some Balance	add-balance.doc	add-balance.pdf
Keeping the Balance When Taking Away	keep-balance.doc	keep-balance.pdf
The Equations Keep Multiplying	keep-multiplying.doc	keep-multiplying.pdf
Equation Writing	equation-writing.doc	equation-writing.pdf
Everything Has a Place	place.doc	place.pdf
Moving Around	moving-around.doc	moving-around.pdf
In a Group	in-a-group.doc	in-a-group.pdf

JPEG Files

Key mathematical images found in the book are also provided on the Teacher Resource CD.

Word Documents of Texts
- Change leveling further for individual students.
- Separate text and images for students who need additional help decoding the text.
- Resize the text for visually impaired students.

Full-Color PDFs of Texts
- Create overhead transparencies or color copies to display on a document projector.
- Project texts on an interactive whiteboard or other screen for whole-class review.
- Post on your website and read texts online.
- Email texts to parents or students at home.

JPEGs of Mathematical Images
- Display as visual support for use with whole class or small-group instruction.

Notes

Notes